A PLUME BOOK

AVERAGE IS OVER

TYLER COWEN is a professor of economics at George Mason University. His blog, *Marginal Revolution*, is one of the world's most influential economics blogs. He also writes for the *New York Times*, the *Financial Times*, and the *Economist* and is the cofounder of Marginal Revolution University. Cowen lives in Fairfax, Virginia.

Praise for Tyler Cowen and *Average Is Over*

"Tyler Cowen's new book, *Average Is Over*, makes an excellent follow-up to his previous work, *The Great Stagnation*, and I expect it will set the intellectual agenda in much the way its predecessor did."
—*Slate*

"A lively and worryingly prophetic read . . . Some of the most talked-about issues in present-day America . . . Observations that are genuinely enlightening, interesting, and underappreciated."
—*The Daily Beast*

"A bracing new book."
—*The Economist*

"His most fun and also his scariest book yet."
—*Forbes*

"[Tyler Cowen] has been one of the most important voices."
—NewYorker.com

D0048388

"The author roves broadly and interestingly to make his case, outlining radical economic transformations that lie in store for us, predicting the rise and fall of cities depending on their capacity to adapt to this machine-driven world, and offering policy prescriptions for preserving American prosperity." —*The Wall Street Journal*

"Thomas Friedman—move over. There's a new guy on the block." —*The Tampa Tribune*

"Eminently readable." —The Brookings Institution

"An economist who is a wonderfully entertaining writer but also a deeply humane thinker." —James Surowiecki, author of *The Wisdom of Crowds*

"[Cowen] ventures where few economists have gone before." —Hugo Lindgren, *New York Magazine*

"Tyler Cowen may very well turn out to be this decade's Thomas Friedman." —Kelly Evans, *The Wall Street Journal*

"[Cowen] will have a profound impact on the way people think about the last thirty years." —Ryan Avent, *The Economist*

"A buckle-your-seatbelts, swiftly moving tour of the new economic landscape." —*Kirkus Reviews*

"Cowen has a rare ability to present fundamental economic questions without all of the complexity and jargon that make many economics books inaccessible to the lay reader." —*The American Interest*

TYLER COWEN

AVERAGE
IS
OVER

Powering America
Beyond the Age of
the Great Stagnation

A PLUME BOOK

PLUME
Published by the Penguin Group
Penguin Group (USA) LLC
375 Hudson Street
New York, New York 10014

USA | Canada | UK | Ireland | Australia | New Zealand | India | South Africa | China
penguin.com
A Penguin Random House Company

First published in the United States of America by Dutton, a member of Penguin
Group (USA) LLC, 2013
First Plume Printing 2014

THE LIBRARY OF CONGRESS HAS CATALOGED THE DUTTON EDITION AS FOLLOWS:
Cowen, Tyler.
Average is over : powering America beyond the age of the great stagnation /
Tyler Cowen.
pages cm
Includes bibliographical references and index.
ISBN 978-0-525-95373-9 (hc.)
ISBN 978-0-14-218111-9 (pbk.)
1. Economic forecasting—United States.   2. United States—Economic
conditions—2009-   3. United States—Economic policy—2009-   I. Title.
HC106.84.C69 2013
330.973—dc23   2013016255

Printed in the United States of America
10 9 8 7 6 5 4 3 2 1

To Natasha

I would bring a hammer.

> —Chess grandmaster Jan Hein Donner,
> when asked what strategy he would use
> against a computer

The title of this book was inspired by a series of Thomas Friedman's columns in The New York Times. *Friedman expanded on this idea in a chapter in his book with Michael Mandelbaum,* That Used to Be Us, *entitled "Average Is Over." I owe a debt of gratitude to their work and thinking on this very important subject. I also recommend to the reader Erik Brynjolfsson and Andrew McAfee's* Race Against the Machine, *a book that came out while I was doing the research and writing on this one. I have benefited considerably from reading their work and from conversations with them.*

Contents

PART I

Welcome to the Hyper-Meritocracy

1

Work and Wages in iWorld

This book is far from all good news. Being young and having no job remains stubbornly common. Wages for young people fortunate enough to get a job have gone down. Inflation-adjusted wages for young high school graduates were 11 percent higher in 2000 than they were more than a decade later, and inflation-adjusted wages of young college graduates (four years only) have fallen by more than 5 percent. Unemployment rates for young college graduates have been running for years now in the neighborhood of 10 percent and underemployment rates near 20 percent. The sorry truth is that a lot of young people are facing diminished job opportunities, even several years after the formal end of the recession in 2009, when the economy began to once again expand after a historic contraction.

Many people are seeing the erosion of their economic futures. The labor market troubles of the young—which you can observe in many countries—are a harbinger of the new world of work to come.

Lacking the right training means being shut out of opportunities like never before.

At the same time, the very top earners, who often have advanced postsecondary degrees, are earning much more. *Average is over* is the catchphrase of our age, and it is likely to apply all the more to our future.

This maxim will apply to the quality of your job, to your earnings, to where you live, to your education and to the education of your children, and maybe even to your most intimate relationships. Marriages, families, businesses, countries, cities, and regions all will see a greater split in material outcomes; namely, they will either rise to the top in terms of quality or make do with unimpressive results.

These trends stem from some fairly basic and hard-to-reverse forces: the increasing productivity of intelligent machines, economic globalization, and the split of modern economies into both very stagnant sectors and some very dynamic sectors. Consider the iPhone. The iPhone is made on a global scale, and it blends computers, the internet, communications, and artificial intelligence in one blockbuster, game-changing innovation. It reflects so many of the things that our contemporary world is good at, indeed great at. Today's iPhone would have been the most powerful computer in the world as recently as 1985. Yet to cite two contrasting sectors, typical air travel doesn't go faster than it did in 1970, and it is not clear our K–12 educational system has much improved.

This imbalance in technological growth will have some surprising implications. For instance, workers more and more will come to be classified into two categories. The key questions will be: Are you good at working with intelligent machines or not? Are your skills a complement to the skills of the computer, or is the computer doing better without you? Worst of all, are you *competing against* the computer? Are computers helping people in China and India compete against you?

If you and your skills are a complement to the computer, your wage and labor market prospects are likely to be cheery. If your skills do not complement the computer, you may want to address that mismatch. Ever more people are starting to fall on one side of the divide or the other. That's why *average is over.*

This insight clarifies many key issues, such as how we should reform our education; where new jobs will come from and why (some) wages might start rising again; which regions will see sky-rocketing real estate prices and which will empty out; why some companies will get smarter and smarter, while others just try to ship product out the door; which human beings will earn a lot more and which workers will move to low-rent areas to make ends meet; and how shopping, dating, and meeting negotiations will all change.

What lies ahead of us will be a very surprising time, and it is likely that new technologies already emerging will lead us out of what I called in a previous book "the great stagnation." It is true that there has been a persistent slowdown in real economic growth in the Western world and Japan, but this book suggests how that might plausibly change. It is not the new technologies per se; it is how some of us will use them.

The technology of intelligent machines may conjure up science fiction visions of rebellious robots or computers that feel and maybe fall in love or proclaim themselves to be gods. The reality of the progress on the ground is based on an integration of capabilities rather than on any one thing that might be described as "artificial intelligence." What is happening is an increase in the ability of machines to substitute for intelligent human labor, whether we wish to call those machines "AI," "software," "smart phones," "superior hardware and storage," "better integrated systems," or any combination of the above. This is the wave that will lift you or that will dump you.

The fascination with technology and the future of work has inspired some important writings, including Martin Ford's classic *The Lights in the Tunnel*, the more recent and excellent eBook *Race Against the Machine* by Erik Brynjolfsson and Andrew McAfee, and Ray Kurzweil's futuristic work on how humans will meld with technology. Debates about mechanization periodically resurface, most prominently in the 1930s and in the 1960s but now once again in our new millennium. *Average Is Over* builds upon these influential works and attempts to go beyond them in terms of detail and breadth. In these pages I paint a vision of a future which at first appears truly strange, but at least to me is also discomfortingly familiar and indeed intuitive. As a blogger and economics writer, I find that the question I receive most often from readers is—by far— something like: "What will the low- and mid-skilled jobs of the future look like?" This question is on everyone's mind with a new urgency but it goes back to David Ricardo and Charles Babbage in the nineteenth century. Ricardo was a leading economist of his time who wrote on "the machinery question," while Babbage was the intellectual father of the modern computer and he—not coincidentally—also wrote on how radical mechanization was going to reshape work.

These questions have reemerged as culturally central because we are at the crux of a technological revolution once again. It's becoming increasingly clear that mechanized intelligence can solve a rapidly expanding repertoire of problems. Solutions began appearing on the margins of the world's interests. Deep Blue, an IBM computer, defeated the then–world champion Garry Kasparov in a chess match in 1997. Watson, a computer program, beat Ken Jennings—the human champion—on *Jeopardy!* in 2010, surpassing most expectations as to how quickly this would happen. Interesting developments, yes, but the technological news is becoming more central to our concerns.

We're on the verge of having computer systems that understand the entirety of human "natural language," a problem that was considered a very tough one only a few years ago. Just talk to Siri on your iPhone and she is likely to understand your voice, give you the right answer, and help you make an appointment. Siri disappoints with its mistakes and frequently obtuse responses, but it—or its competitors—will improve rapidly with more data and with assistance from crowd-sourced recommendations and improvements. We're close to the point where the available knowledge at the hands of the individual, for questions that can be posed clearly and articulately, is not so far from the knowledge of the entire world. Whether it is through Siri, Google, or Wikipedia, there is now almost always a way to ask and—more importantly—a way to receive the answer in relatively digestible form.

It must be emphasized that every time you use Google you are relying on machine intelligence. Every time Facebook recommends a new friend for you or sends an ad your way. Every time you use GPS to find your way to a party.

Don't write off those robots either, even if they may never pray to God or pass for human beings. In 2011 Taiwan-based Foxconn, the world's largest contract electronics manufacturer, announced a plan to increase the use of robots in its factories one hundredfold within three years, bringing the total to one million robots. After recent wage increases in China—to levels still low by Western standards—the company doesn't consider its labor so cheap anymore. In the United States as well, the use of industrial robots is booming, and the likely future for North America is that of a coherent economic unit where the United States, Canada, and Mexico band together to make major investments in customized robot production and then use these investments to dominate global manufacturing.

Robot-guided mechanical arms are common in the operating room, and computers spend more time flying our planes than do the pilots. South Korea is experimenting with robotic prison wardens that patrol when the inmates do something wrong and report the misdeeds.

Driverless cars are already operating on the streets of Berlin and Nevada, and Florida and California have passed bills to legalize computer-commanded "driverless cars" on their roads. Google's team has test-driven hundreds of thousands of miles with these cars, so far without an accident or major incident; the one reported five-car pileup happened after a human took over from the computer. Some Google employees have their self-driving vehicles take them to work. These car robots don't look like something from *The Jetsons*; the driverless features on these cars are a bunch of sensors, wires, and software. This technology *works*.

There is now a joke that "a modern textile mill employs only a man and a dog—the man to feed the dog, and the dog to keep the man away from the machines."

Software is also encroaching upon journalism. One experiment found that the intelligent mechanized analysis of Narrative Science, a start-up from Illinois, can do a passable job of taking statistics and writing up descriptions of sporting events, company financial reports, and macroeconomic data. These programs won't soon be at the frontier of creative journalism, but they may soon be generating a lot of run-of-the-mill news for purposes of search and storage. They also may take away some jobs: Should the local newspaper really send a reporter down to that minor league baseball game? Software is not only taking a shot at writing essays but also grading them and providing instant feedback on student work in progress, analysis that is well beyond grading multiple-choice quizzes. These programs still need to work out some bugs (a clever student can game them with coherent-sounding nonsense), but they are much further along than we had been expecting five or ten years ago. Writers and teachers need to consider what aspects of their work are better done by intelligent-machine analysis and look closely at the irreplaceable value they do provide.

Date-matching algorithms are steering our love lives and replacing the matchmaker. Match.com recently improved its services, and as of summer 2011 more than half of the emails sent on the service originate from recommended matches, rather than from unaided individual choices. Better algorithms often are seen as the future of the sector, whether or not they really find the best person for us. Arguably the machine recommendations are a way of tricking the user into making a plausible date choice rather than cruising more profiles and postponing a decision; that possibility illustrates our willingness to defer to the machines, even when they aren't necessarily better at the task at hand.

On Netflix, it is now standard for users to consult or follow the system's algorithm for movie choices. We make the choice, but we have a new smart partner when it comes to movie watching.

Some of these developments may end up creeping us out, precisely because they might prove effective. The city of Santa Cruz,

California, has already begun using mechanized intelligence to deploy police officers against car and property theft. The program, written by a team of social scientists and two mathematicians, generates predictions about which areas and windows of time are most likely to see property crimes; the model has been published in *Journal of the American Statistical Association*. As new crimes occur, the predictions are recalibrated daily; it's based on some of the models for predicting aftershocks from earthquakes. The program awaits further study, but this will not be the last attempt to automate crime prevention. The TSA is experimenting with software that tries to detect, by scanning body language, which plane passengers have hostile intentions.

Not each and every one of these innovations will pay off. But let's ask a few questions. First, in which major areas do we see ongoing technological advances exceeding expectations from just a few years ago? Second, in which areas do we see a lot of new and promising technological works in progress? Third, in which areas can we expect the general forces propelling innovation (say globalization or Moore's law that processing power for computers will continue rising at rapid rates) to remain powerful? Finally, can we see evidence that these areas are already influencing economic statistics measuring our nation's well-being? I'll get into more detail on all of these questions, but for now the point is that the areas of the economy identified in the answers to these questions all overlap on one technology: mechanized intelligence. And its effect on economic statistics is trending up.

The next step perhaps is that we will start to see just how well some of these machines can predict our behavior. One of Isaac Asimov's most profound works is his neglected short story

Franchise. In this tale democratic elections have become nearly obsolete. Intelligent machines absorb most of the current information about economic and political conditions and estimate which candidate is going to win. (In fact a small number of variables, such as the change in GDP, the unemployment rate, the inflation rate, and the presence of a major war, predict presidential elections pretty well.) In the story, however, the machines can't quite do the job on their own, as there are some ineffable social influences the machines cannot measure and evaluate. The American government thus picks out one "typical" person from the electorate and asks him or her some questions about moods. The answers, combined with the initial computer diagnosis, suffice to settle the election. No one needs to actually vote.

This will sound outrageous to many. It seems to cross a precious line of liberty and freedom. But perhaps we are not as free as we might think in the first place. Given your background, your friends, your family, the books you read, and the movies you watch, how surprising is your vote in a federal election? The future of technology is likely to illuminate the unsettling implications of how predictable we are and indeed in 2012 political campaigns invested heavily in predicting where to find supporters and important swing districts.

Here is a short paragraph from *The New York Times* that illustrates where we already have arrived:

> As Pole's computers crawled through the data, he was able to identify about 25 products that, when analyzed together, allowed him to assign each shopper a "pregnancy prediction" score. More important, he could also

estimate her due date to within a small window, so Target could send coupons timed to very specific stages of her pregnancy.

The computers crawling over this mountain of records of individual retail sales used an algorithm that recognized pregnant women buying lots of calcium, magnesium, and zinc supplements early in their pregnancies, unscented lotion around the beginning of the second trimester, and hand sanitizer and extra-big bags of cotton balls as they approach the date of delivery.

Whether we like it or not, our sparring partners will use mechanized intelligence during their business contests. When a major negotiation is being conducted, or when potential business partners are being introduced, the interaction will be recorded, processed, and analyzed in real time, just as the genius machine Watson parses a *Jeopardy!* question. Each party to the communications might receive a real-time report on when the other people are likely lying, what their stress levels are, how detailed their narratives are, who in the group really speaks with authority, and how many first-person pronouns they are using, all based on an analysis of voice data. From this data, and other measurable factors, a program will construct and transmit some kind of "read" on the conversation. The machine doesn't have to reach perfection or even come close, it just has to do better than you.

There is work in progress on using software to detect human lies, based on computer analyses of our voices; Dan Jurafsky of Stanford University and Julia Hirschberg of Columbia University are two leaders in this area. They claim their programs already can detect deception more readily than can human observers; in any case, improvements and refinements stand a good chance of getting us there.

Imagine a vibrating iPhone in one's pocket transmitting signals, based on the computer analysis—a slight vibration indicating each time a lie is told. Or a message could appear in your contact lenses. But it isn't just about the gadgetry. Eventually it will be commonly understood that such analyses are going on in real time. Negotiators will be trained to fool or otherwise throw off the voice-tracking programs. In turn the programs will be improved to keep up with these tactics, setting up a never-ending "arms race" between technologies of deception and detection. And a new kind of sophisticated social interaction will develop. That is bigger news that any new gadget.

Will we see this social interaction prove effective in business negotiations and stop there? Our human curiosity is likely to run its usual courses. There will be new options for calming nerves and doubts on first dates: Does she like me? Is this guy trying to hit on me? Is he actually married? Will she let me kiss her on the cheek? Of course, we won't be able to stop other people from bringing inconspicuous recording and analyzing devices to our face-to-face meetings. Someone will surely try to develop a way to measure the genetic information of the people one is dealing with, as was portrayed in the seminal movie *Gattaca*.

And then there will be mechanized intelligence in our home. Imagine recording and analyzing scenes from your living room and bedroom. The very idea might be loathsome, and it is probably not useful to see a running prediction on the likelihood of mom and dad staying together as a couple. But could you avoid the temptation to take a peek at that number every now and then?

We may tend to think of mechanized intelligent analysis as primarily useful in judging other people, but it will also have the potential to promote self-knowledge. During a date, a woman might consult a pocket device in the ladies' room that tells her how much she really likes the guy. The machine could register her pulse, breathing, tone of voice, the level of detail in her narrative, or whichever biological features prove to have predictive power.

Self-scrutiny doesn't have to be restricted to matters of the heart. Which products do we really like or really notice? How are we responding to advertisements when we see them? Consult your pocket device. There is currently a DARPA (Defense Advanced Research Projects Agency, part of the Department of Defense) project called "Cortically Coupled Computer Vision." The initial applications help analysts scan satellite photos or help a soldier-driver navigate dangerous terrain in a Jeep. Basically, the individual wears some headgear and the device measures neural signals whenever the individual experiences a particular kind of subconscious alert (Danger! or, Salt! or, Familiar!). Someday we will be able to dispense with the headgear or otherwise make the device less burdensome and less conspicuous. You will be able to walk down the aisle of a shopping mall or scan a window in a store, and the device will record and note whichever products grabbed your attention. You could program it to prompt you at the time or to store the information, with appropriate tags, for future reference, perhaps combined with a web search for appropriate coupons.

If you're not interested, the store probably will be. Your shopping cart will use GPS to track your moves through the store, including which aisles you visit most often. Or they can track you by video and smart cameras, using face recognition technology to aggregate data across your different store visits. Stores have even started to use cameras on test subjects to track shoppers' retinas, to see how and when they notice products, or to lock on to cell phone signals and track and analyze shoppers' movements. When you enter the store, they will promise you a discount, but only if you run your card through the cart and identify yourself. Many will grab the discount, just as they participate in frequent-shopper programs, even though it means the food company knows what they are buying. They'd rather have the lower prices than the privacy—as indeed would I.

There are already pilot projects along these lines. Companies have discovered that if the first product seems cheaper than expected, the customer is more trusting, and more willing to spend, for the duration of the shopping trip. So if the intelligent machines spot you coming in the door and heading to the gourmet chocolate, maybe there will be a quite temporary sale, communicated by electronic sensors—for your benefit, of course.

It's easy to see how comparable developments, as applied to analyzing human behavior and intentions, could thwart a lot of relationships or business deals in the works. The sorry truth is that if we knew all or even some of the bad things about our prospective partners, we might be so cautious that we never take a romantic leap. As it stands, the world is set up to overreact to negative information, as even a whiff of scandal causes us to lose trust in other individuals. We will need some significant cultural changes to make do with an increase in the "warts and all" coverage that mechanized intelligent analysis will soon be delivering about virtually all notable public figures and many private individuals too. What if a woman's pocket device reported during the bathroom break that her date smiled at the waitress for a little too long? Sometimes the guy is a creep who should be dumped, but, well, in this particular case, maybe not. The positive illusions (all our children are above average) that help get us through everyday life could so easily wither in the face of sangfroid machine critiques. Not this year or next year, but most likely within our lifetimes.

Smart software is already used to ferret out phony web reviews on sites such as Amazon and TripAdvisor. A team of Cornell researchers discovered how to spot phony reviews about 90 percent of the time. The paid-for fake reviews use too many superlatives, too many vague evaluations, and not enough detailed descriptions, and they also use the words "I" and "me" more often, perhaps as a substitute for knowing what they are talking about. In due time the fake reviews will evolve and will be harder to spot by these older methods, and the fake-spotting software will have to catch up—another ongoing arms race.

Similarly, researchers are working on how to spot the liars in online dating, so we can expect software to follow suit. Preliminary results suggest that liar profiles avoid the use of the word "I" (the opposite of liar user reviews), use lots of negation (they write "not sad" rather than "happy"), and they write shorter self-descriptions, presumably to avoid having to manage and keep all the lies consistent. Profile writers who are lying about their age or weight tend to devote more space to boasting about their personal achievements. Expect improvement in our abilities to nail the cheaters.

It is no accident that we are seeing so many signs of significant progress in mechanized intelligent analysis, albeit in varying stages of maturity. First, Moore's law about ongoing advances in processing speed has continued to pay off, with no immediate end in sight. Second, the machine intelligence sector is largely unregulated. If you compare it to health care as a world-altering, stagnation-ending breakthrough industry, regulatory obstacles are a far greater problem for pharmaceutical companies and for hospitals than for the likes of Google, Amazon, and Apple. Health care, with its physician licensing, Byzantine hospital regulations, and FDA approval process, also makes most of its changes quite slowly, for better or worse. It's not just about the laws, but also because doctors and patients often have very conservative or moralistic views about how health care should be done; just look at the recent controversies over stem cell treatments and genetic engineering. Health care is an ethical minefield and arguably we *should* be especially cautious when evaluating new medical and institutional breakthroughs. In any case, we can expect slower progress.

When it comes to mechanized intelligent analysis, patent law can be a problem but for the most part the paths forward are relatively free of regulatory obstacles. Some applications, such as driverless cars, do face potential lawsuits; for instance, imagine the first time an out-of-control driverless car runs down a child. The parents would likely sue the deep-pocketed company that made the car, in addition to suing the owner of the car and maybe also the person who programmed the software of the car. Given this likelihood, there may be delays in turning the cars into marketable products. Maybe it will happen first in some part of the world far less litigious than the United States.

Still, very often entrepreneurs and scientists can do the work behind smarter machines, and develop usable products, without needing much special permission from the powers that be. Deep Blue and Watson and Siri were brought to the world without facing many regulatory or legal barriers. Technological progress slows down when there are too many people who have the right to say no, but software in general gets around a lot of the traditional veto

points. The key work is done in the individual mind and with relatively small teams and in computer code, and it's hard to hold back
the innovations by law or custom.

So how does this new world change what young people should
expect from their work lives?

2

The Big Earners and
the Big Losers

To put the question in the bluntest possible way, let's say that machine intelligence helps us make a lot more things more cheaply, as indeed it is doing. Where will most of the benefits go? In accord with economic reasoning, they will go to that which is scarce. In today's global economy here is what is scarce:

1. Quality land and natural resources
2. Intellectual property, or good ideas about what should be produced
3. Quality labor with unique skills

Here is what is not scarce these days:

1. Unskilled labor, as more countries join the global economy
2. Money in the bank or held in government securities, which you can think of as simple capital, not attached

to any special ownership rights (we know there is a lot
of it because it has been earning zero or negative real
rates of return)

We see high returns to resource owners, such as the new-
resource millionaires and billionaires of Brazil, Russia, Canada,
and Australia, and similarly huge revenues for intellectual property
giants such as Apple and other innovators in that sector. The current array of scarce and plentiful resources now means high wages
or capital gains to talented and inventive workers, and pretty low
returns on ordinary labor and ordinary savings. Don't forget this
larger picture, but in this book I'll be focusing on the labor market
side of the equation and what that will mean in our lifetimes.

It was true in the great Industrial Revolution of the nineteenth
century and it is true now: Machines do not put us all out of work,
as eventually machines create new jobs just as they destroy old
ones. It is also true that the new machines of our age will give rise
to new and different workplaces and create a new set of winners and
losers.

General Philip M. Breedlove, US Air Force vice chief of staff,
works with military drones. He recently remarked, "Our number
one manning problem in the air force is manning our unmanned
platforms." This includes workers to fix and maintain the drones
and analysts to sort through the subsequent video and surveillance
feeds.

According to the air force, keeping an unmanned Predator
drone in the air for twenty-four hours requires about 168 workers
laboring in the background. A larger drone, such as the Global
Hawk surveillance drone, needs about 300 people working in the
background to make the mission feasible. To compare, the

operation of an F-16 fighter aircraft requires fewer than 100 people for a single mission.

As intelligent-analysis machines become more powerful and more commonplace, the most obvious and direct beneficiaries will be the humans who are adept at working with computers and with related devices for communications and information processing. If a laborer can augment the value of a major tech improvement by even a small bit, she will likely earn well.

That means humans with strong math and analytic skills, humans who are comfortable working with computers because they understand their operation, and humans who intuitively grasp how computers can be used for marketing and for other non-techie tasks. It's not just about programming skills; it is also often about developing the hardware connected with software, understanding what kind of internet ads connect with their human viewers, or understanding what shape and color makes an iPhone attractive in a given market. Computer nerds will indeed do well, but not everyone will have to become a computer nerd.

This dynamic—higher earnings for those who "get" computers— affects many sectors beyond Silicon Valley. This isn't merely a story about science, technology, engineering, and math majors (STEM), because a lot of scientists aren't getting jobs today, especially if they don't do the "right" kind of science. Does anyone envy the job prospects of a typical newly minted astronomy PhD? On the other hand, Mark Zuckerberg of Facebook fame was a psychology major, and insights from psychology helped him make Facebook into a more appealing and alluring site. The ability to mix technical knowledge with solving real-world problems is the key, not sheer number-crunching or programming for its own sake. Number-crunching skills will be turned over to the machines sooner or later.

Marketing

Despite all the talk about STEM fields, I see *marketing* as the seminal sector for our future economy.

A salesperson can use knowledge of computers or engineering to sell a complex technical product to a technically sophisticated user, and in fact such knowledge might these days be required to sell effectively. That's based in some STEM skills, but it's also marketing. The entertainment sector is about marketing, especially as the internet makes the array of available cultural products ever more crowded. It might appear that a masseuse is not much affected by computers, at least provided you are skeptical about these robots that now offer massages. Nonetheless, masseuses increasingly market themselves on Google and the internet. These masseuses fit the basic model that favors people who can blend computer expertise with an understanding of how to communicate with other people. Again, it is about blending the cognitive strengths of humans and computers.

We can expect a lot of job growth in personal services, even if those jobs do not rely very directly on computing power. The more that the high earners pull in, the more people will compete to serve them, sometimes for high wages and sometimes for low wages. This will mean maids, chauffeurs, and gardeners for the high earners, but a lot of the service jobs won't fall under the service category as traditionally construed. They can be thought of as "creating the customer experience." Have you ever walked into a restaurant and been greeted by a friendly hostess, and noticed she was very attractive? Have you ever had an assistant bring you coffee before a meeting, touching you on the shoulder before leaving the cup? Have you gone to negotiate a major business deal and been greeted by a mass of smiles and offers of future friendship and collaboration? All of those people are working to make you feel better. They are working at marketing.

It sounds a little silly, but making high earners feel better in just about every part of their lives will be a major source of job growth in the future. At some point it is hard to sell more physical stuff to high earners, yet there is usually just a bit more room to make them feel better. Better about the world. Better about themselves. Better about what they have achieved.

The growing importance of marketing integrates two seemingly unrelated features of the modern world: income inequality and increasing pressures on our attention. The more that earnings rise at the upper end of the distribution, the more competition there will be for the attention of the high earners and thus the greater the importance of marketing.

If you imagine two wealthy billionaire peers sitting down for lunch, their demands for the attention of the other tend to be roughly equal. After all, each always has a billion dollars (or more) to spend and they don't need to court each other for favors so much. There is a (rough) parity of attention offered and received. Of course, some billionaires are more important than others, or one billionaire may court another for the purpose of becoming a megabillionaire, but let's set that aside.

Compare it to one of those same billionaires riding in a limousine, with open windows, through the streets of Calcutta. A lot of beggars will be competing for the attention of that billionaire, and yet probably the billionaire won't much need the attention of the beggars. The billionaire may feel overwhelmed by all of these demands, and yet each of these beggars will be trying to find some way to break through and capture but a moment of the billionaire's attention.

This in short is what the contemporary world is like, except the billionaire is the broader class of high earners and the beggars are wealthier than in India. Instead of begging, there is a large class of people trying to command our attention using modern technologies such as email, spam, AI-targeted advertisements, coupons, Groupons, direct mail, advertising supplements in your credit card bill, and flashing ads on the internet, among hundreds of other techniques. All will appeal to our vanity and promise us a splendid customer experience of some kind or another.

For the high earners, life will feel better than ever before, but at the same time life will feel more harried and more overloaded with information than ever before. These phenomena are in fact two sides of the same coin, and this tends to get overlooked. After all, when income and wealth disparities are pronounced, everyone who isn't at the very top will be scrambling for the attention of those who are.

This is not to say what is being marketed to the highest earners is worthless or necessarily corrupt. The 19 percent below the 1 percent at the top do not depend for their wages or earnings purely on getting or giving the right kind of attention. But getting attention will continue to be a critical function in the new world of work and is likely to require ever-greater effort and sophistication.

There are other essential roles to be played, of course. For the top 1 percent of earners in America, a lot of the big gains come in the zip codes of New York City, the Upper West Side zip code of 10023 being number one. The Upper East Side does well, as does Scarsdale (a suburb of New York), Cupertino, and Potomac and Bethesda near Washington, D.C. This reflects how many of the very highest earners come from internet companies and finance, with some government and law thrown into the mix. There has been Steve Jobs, Mark Zuckerberg, and Bill Gates, but finance is a common source of riches within the very highest tier of earners. To give one extreme but illuminating example, in 2007 the top twenty-five hedge fund earners pulled in more income than all the CEOs of the S&P 500 put together. The modern financial sector has computers, computerized trading, arbitrage, super-rapid communications, and computerized risk assessment at its core.

But don't just focus on those computers; it's also about management. The CEOs and higher-level managers are paid handsomely to assemble and direct the individuals who work every day with mechanized intelligent analysis. If you have an unusual ability to spot, recruit, and direct those who work well with computers, even if you don't work well with computers yourself, the contemporary world will make you rich. If we look at the increase in the share of income going to the top tenth of a percent from 1979 to 2005, executives, managers, supervisors, and financial professionals captured 70 percent of those gains.

Another development is this: The better the world is at measuring value, the more demanding a lot of career paths will become. That is why I say "Welcome to the hyper-meritocracy" with a touch of irony. Firms and employers and monitors will be able to measure economic value with a sometimes oppressive precision.

The coming world of hyper-meritocracy I'm sketching is not necessarily a good and just way for an economy to run. It will be more productive, and it is true that economic productivity is correlated with many good or apparently good qualities, including higher earnings, better health (usually), and greater financial responsibility toward one's family. But are productive people always happier? Always more creative and reflective? Do they always bring more joy to others? Must we all try to be highly productive, no matter how ill-fitting the suit may be? I am not trying to settle these debates at any kind of moral level, but rather to ask how very productive people will fare in the job market as we move forward. We'll be reading of some particular good or bad results for the productive and the less productive, but let's investigate where those results are coming from before we get emotionally involved in an overall moral judgment.

Note that much the same could be said regarding my numerous references to the words *smart* and *intelligent* and *conscientious* and *talented*. Sure, these are good qualities overall, but we don't always need to prize them above all other human qualities. How we value them does have a moral implication, but such moral judgments can be left to each and every one of us to make.

In any case, the slacker twenty-two-year-old with a BA in English, even from a good school, no longer has such a clear path to an upper-middle-class lifestyle. At the same time, Facebook, Google, and Zynga are now so desperate for talent that they will buy out other companies, not for their products, but rather to keep their employees. It's easier and cheaper to buy the companies than to try to replicate their recruiting or lure away their best employees. Often the purchased product lines are abandoned. A recent report laid out how these acquisitions work: "'Engineers are worth half a million to one million,' said Vaughan Smith, Facebook's VP of corporate development, who has helped negotiate many of the 20 or so talent acquisitions made by Facebook in the last four years." The technology blogs call this being "acqhired," and this practice is being ramped up in what is otherwise a slow job market. It's not slow for those who work with the intelligent machines.

Below the mega-achievers and the billionaires, job growth for ordinary workers is consistent with this story of a bifurcated job market. Measured in the fall of 2011, over the span of a year jobs went up in electronic shopping establishments (up 11 percent), up in the category of "internet publishing, broadcasting, and web search portal" (up 20 percent), and up in "computer systems design, programming, and related," (up 5 percent). Professions that depend on STEM are the second-fastest-growing group in the United States, another example of the strong demand for technical skills.

Management

Whether you work for Google or for McDonald's, many ordinary wage earners are already asking why managers deserve so much more money than those lower down on the ladder. To better understand why the new world of work is turning out to be so good for qualified managers, let's consider this personal perspective.

As a professor, I am given a research assistant each year. Over the last twenty or so years, I have received some extraordinary assistance from some very good workers, students, and eventually, peers and coauthors.

About once a year I receive an offer, usually by email, from someone who wants to work as my research assistant for free. Typically the offer is accompanied by a resume, and for the most part these resumes appear quite good. The emails sound reasonable and friendly.

I turn such offers down. I don't think the applicants are lemons, but still I find that one research assistant is for me the right number, at least if I have a good one, as is usually the case. Even when it comes to the assistant whom I have the time to manage, I am most of all concerned about having a conscientious person at my disposal.

The work with an RA is basically a team relationship, and the core problem is that I don't have the time to build another team, even if it doesn't cost me any money upfront. I don't have the time to work with and manage another person. To put this point in a broader business context, until another good manager is hired, there is no point in employing another assistant. It's the manager who is the scarce input, and that is one way to think about why managerial salaries have been going up so much. Managers play a role of growing importance in coordinating complex, large-scale production processes.

The intelligent-machine revolution is making the modern workplace more team-oriented. Machines, computers, and the internet allow us to string together large teams of cooperating labor, sometimes from around the world. When Apple produces an iPad, it strings together a network of producers in what is a virtually miraculous pattern of economic cooperation, ranging from designers in Cupertino, California, to Toshiba hard drives (a Japanese company, producing in the Philippines and China) to computer chips from Taiwan, to final assembly in China and marketing and retail expertise from around the United States.

The teamwork required to divide up a larger project into pieces requiring different skills in order to produce a better result has been a critical part of successful economies since Plato and Xenophon started talking about it thousands of years ago. But things have changed a little. Adam Smith wrote in 1776 that "division of labor is limited by the extent of the market" and since that time the market has gotten a whole lot bigger. The division of labor has expanded

accordingly, so skills get more and more finely focused and tuned—
and that means the screwups of a single person can damage a very
large and very valuable production chain. To hire a risky and iffy
worker, without a competent overseer, simply isn't worth it, no matter how low the wage. And so a lot of workers have a hard time being picked up and integrated into productive teams.

It is precisely that process that managers are paid to make work
more efficiently. It is a process that is continuing its long, long trend
toward increasing importance. And, finally, it is why managers are
being paid more.

Workers

Many of our service sectors, a lot of our precision manufacturing,
our healthcare and education sectors, government bureaucratic employment, and our creative industries—to name but a few
examples—rely less and less on the brute force of additional human
labor. Even the military is more about manipulating advanced technology than just aiming a gun and shooting it, or stabbing with a
bayonet. One bad soldier or engineer can ruin the efforts of many
others, or if someone programs the drone wrong, all sorts of problems can arise. When it comes to these complex tasks, people have
to know what they are doing, they have to want to learn, and they
have to want to cooperate with their fellow workers. That means a
growing role for workplace morale and consistency of execution.

Team production makes the quality of "conscientiousness" a
more important quality in laborers. Managers need workers who are
reliable. If you have a team of five, one unreliable worker is wrecking

the work of four others. If you have a team of twenty-five, one unreliable worker can negate the work efforts of twenty-four others. Managers will stay away from possibly destructive labor and they will put a lot of care into building and maintaining their teams.

It's not just that the bad workers are lazy or maybe destructive. It's that low quality workers spread bad morale to many others in the building. It's the troublemakers in a workplace who usually end up getting everyone else at one another's throats. Contemporary employers really do want to get these people out of the company altogether, or performing distant, solo jobs, such as driving a truck from one warehouse to another, rather than standing around the watercooler.

The growing value of conscientiousness in the workplace helps women do better than men at work and in colleges and universities. At my daughter's recent college graduation ceremony the awards for the top achievers in all of the school's programs and departments went almost entirely to women, including awards in science and mathematics.

It is well known from personality psychology, and confirmed by experience, that women are on average more conscientious than men. They are more likely to follow instructions and orders with exactness and without resentment. That means better jobs and higher wages for a lot of women in this new world of work, without a comparable upgrade for a lot of the men. There is plenty of evidence that women are less interested in direct workplace competition and more likely to work well in teams and more likely to seek work in teams. You can think of men as the "higher variance" performers at work. That means some men are more likely to be the very highest earners and also to exhibit extreme dedication to the task, perhaps to the point of being monomaniacal. At the very top there will be a disproportionate share of men as CEOs, top chefs, and also chess players, among many other avocations. Other men, in greater numbers, will be more irresponsible, more likely to show up drunk, more likely to end up in prison, and more likely to become irreparably unemployable.

When the order and coherence and reliability of the supply chain are especially important factors, whom are you going to choose for those middling-level jobs? We already see that young men are the group hit hardest by the recession, in terms of their labor market opportunities. For instance, the unemployment rate for males aged 20–24 has averaged 14.16 percent for 2012.

Here is another, more general way to think about the shifting gender balance of power in education and parts of the workplace. The wealthier we become, the greater a cushion we have against total failure, starvation, and other completely unacceptable outcomes. In such a world, both women and men will indulge some propensities that otherwise might be stifled or kept under wraps or that would not have been affordable fifty or one hundred years ago. For some men, these propensities are quite destructive and this turns them into labor market failures.

Conscientiousness is especially valuable in two other important parts of the labor market: health care and personal services. Most healthcare workers are not doctors, and many of them are not geniuses. Nonetheless, you want to make sure these workers wash their hands when necessary, write down the correct information on the patient's chart, and measure the lab quantities correctly. Again, that's conscientiousness, and due to population aging, the number of healthcare jobs will continue to grow. It is no accident that female workers are especially well represented in these fields, as they are in education.

As workers are displaced by smart machines in manufacturing and other areas, more individuals will be employed as personal trainers, valets, private tutors, drivers, babysitters, interior designers, carpenters, and other forms of direct personal services. These are all areas where a patron—often a family or individual—expects a commission or request to be followed. "Pick up my kid from school." "Fix the electrical wiring." "Show up for my lesson at six o'clock." Most of these jobs require some applied skills but not monomaniacal commitment at the highest levels of world-class achievement. The premium is on conscientiousness, namely whether the worker can follow some straightforward requests with extreme reliability and basic competence. If you are looking to hire a concierge butler, the person really does have to be trustworthy.

If you're a young male hothead who just can't follow orders, and you have your own ideas about how everything should be done, you're probably going to have an ever-tougher time in the labor markets of the future. There won't be much room for a "rebel without a cause" or, for that matter, a rebel *with* a cause. It's not surprising that teen employment has been falling since the 1990s, well before the recent recession.

Let's draw up a simple list of some important characteristics in technologically advanced modern workplaces:

1. Exactness of execution becomes more important relative to an accumulated mass of brute force.
2. Consistent coordination over time is a significant advantage.
3. Morale is extremely important to motivate production and cooperation.

Recent research bears out these principles. Economists Timothy F. Bresnahan, Erik Brynjolfsson, and Lorin M. Hitt performed an extensive poll of managers, combined with follow-up interviews. They found that in the opinions of managers, computer use increases the need for skilled workers, computers tend to increase workers' autonomy, and computers increase the need and ability for management to monitor their workers. All of those features will feed into the need for workers who are smarter, better trained, and more conscientious.

The days of a lone worker in the field pushing a hoe are over, at least as a way to feed families. Think of the public works projects of the 1930s, such as paving a road. A healthy worker always can add some brute force to the endeavor, for instance by carrying bricks from one place to another on the construction site. The workers don't have to be brilliant—they require only a minimum of training—and while conscientiousness plays a role, the monitoring and enforcement problems are relatively straightforward, as the workers either carry the bricks or they do not.

To continue the contrast between two ways of organizing the world of work, I find it useful to go back in time to an era when income inequality was high and many individuals did not have adequate training for higher-wage jobs. From such a setting, I am struck by a passage by Henry Mayhew, who wrote about London labor markets in the mid-nineteenth century:

> Among the wares sold by the boys and girls of the streets are:—money-bags, lucifer-match-boxes, leather straps, belts, firewood . . . fly-papers, a variety of fruits, especially nuts, oranges, and apples; onions, radishes,

water-cresses, cut flowers and lavender (mostly sold by girls), sweet-brier, India rubber, garters, and other little articles of the same material, including elastic rings to encircle rolls of paper-music, toys of the smaller kinds, cakes, steel pens, and penholders with glass handles, exhibition-medals and cards, gelatine cards, glass and other cheap seals, brass watch-guards, chains, and rings; small tin-ware, nutmeg-graters, and other articles of a similar description, such as are easily portable; iron skewers, fuzees [matches], shirt-buttons, boot and stay-laces, pins (and more rarely needles), cotton bobbins, Christmasing (holly and other evergreens at Christmas-tide), May-flowers, coat-studs, toy-pottery, blackberries, groundsel, and chickweed, and clothes'-pegs.

Grabbing the attention of customers right there on the grubby street was critical to this life. Marketing was primitive at best. And it is striking how little overhead most of these jobs had. No benefits. No rent. Not much advertising, at least not beyond standing on the corner and screaming out the name of the product. No human resources department and no real risk of lawsuits. There was nothing romantic about these bleak and poorly paid jobs, but they are an interesting contrast to the way employment works today.

For a suitable contrast, consider the office suite at Google. The workers receive health insurance, extensive training, a lot of time and attention, and an attractively decorated office, including a dedicated play space with fun toys. The food in the cafeteria is excellent and includes Indian curries, finely spiced, and tasty vegan and vegetarian choices. These jobs have overhead.

Not just anyone can show up and grab a desk, because the

company cares very much about who is part of the team. To work at Google you have to go through a tough interview process. Here are three interview questions—typical of the rigor—that have been asked of prospective Google employees:

"How many times a day does a clock's hands overlap?"

"You need to figure out the highest floor of a one-hundred-story building an egg can be dropped from without breaking. The question is how many drops you need to make. You are allowed to break two eggs in the process."

"The probability of a car passing a certain intersection in a twenty-minute window is 0.9. What is the probability of a car passing the intersection in a five-minute window?"

Easy enough? There's a whole book titled *Are You Smart Enough to Work at Google?* by William Poundstone. A few minutes reading it will make the answer clear to most readers, even if the word *smart* is not exactly the right word (Picasso was a genius but I doubt he could have landed a job at Google's Mountain View headquarters).

It's a far cry from the London city streets, where any kid could show up and sell gelatin cards, taking his chances with market demand.

Might a boss wish to mix the two modes of employment? Google could keep its current full-timers at their desks and play stations, and rope off a part of the building where just about anyone could come in and try to sell something. You would find venture entre-preneurs peddling their new ideas to Google staff and immigrants selling lunch tacos. Imagine a floor of the building devoted to such a commercial free-for-all.

Google doesn't mix modes of employment this way because it

wishes to control what goes on under the Google aegis. Mixing modes would involve a lot of distractions, it would blur what Google is all about, it would require enforcement measures to keep the traditional Google-linked parts of the building orderly, and sooner or later it would bring a lot of lawsuits.

At a lot of growing firms, the capital surrounding a worker is going up. Benefits cost more, the firm cares more about its morale and workplace environment, the firm cares more about its overall reputation (workers represent a firm to the broader outside world), and the firm faces a higher risk of lawsuits. Workers are more capable of doing damage to a firm than in times past, so companies are often getting choosier whom they hire. It is easier to destroy than create, and the more valuable and the more precision-based that firms become, the more they will worry about destruction of value coming from workers.

Any time there is a discussion of management strategies, you probably will hear a lot of words like *teamwork, morale,* and *integrity.* That's all well and fine, but what if we substitute *exclusion* for all those nice warm phrases? They would be the same management strategies merely explained from a different point of view, namely of those who are kept away. There is no high morale without exclusion, no integrity without exclusion, and no corporate culture without exclusion. If the management styles at today's quality companies seem so nice, so friendly, and sometimes so downright heartwarming, it is possible only because those cultures are so very picky, snobbish, and elitist at the same time. There is no open door.

Basically what's happening is that a lot of jobs are becoming more like Google—you have to meet a certain grade or you are out—and there is another lower tier of jobs becoming just a bit more like Mayhew's 1850 London streets, albeit at higher wage levels. Structural unemployment is the bumpiness that we experience when some individuals have to move from one mode of work to the other, because not everyone is suited to work at Google.

You might think it's only Google and a few elite firms moving in this direction—meet a certain grade or you are out—but the practice is spreading to many corners of the job market. For instance, it's now common that a fire chief has to have a master's degree. That may sound silly and perhaps you think a master's degree has not very much to do with putting out fires. Still, often it is desired that a firefighter be trained in emergency medical services, anti-terrorism practices, and fire science (for instance, putting out industrial fires), and there is a demand for firemen who, as they move into leadership roles, can do public speaking, interact with the community, and write grant proposals. A master's degree is no guarantee of skill in these areas, but suddenly the new requirements don't sound so crazy anymore.

In fact, there is a whole host of jobs that now very often require college degrees, although of course not usually a master's. These jobs include dental lab technicians, chemical equipment operators, medical equipment preparers, and buyers and purchasing agents, among others.

We have been seeing what is called "labor market polarization," a concept that is most closely identified with MIT labor economist David Autor. Labor market polarization means that workers are, to an increasing degree, falling into two camps. They either do very well in labor markets or they don't do well at all. It's hardly the case that America has lost its middle class as of 2013, and I would urge you to stay away from some exaggerated accounts of the middle class having been "decimated," but looking toward the future the trend is clear: The middle of the distribution is thinning out and this process appears to have a long ways to run. And to be blunt— while I know I can't prove this—I wonder how much of the middle class consists of people in government or protected service-sector jobs who don't actually produce nearly as much as their pay.

Of the jobs lost during the recession, about 60 percent of them were in what are called "mid-wage" occupations. What about the jobs added since the end of the recession? Seventy-three percent of them have been in lower-wage occupations, defined as $13.52 an hour or less. This general trend, namely more rapid growth in low-paying jobs, can be seen in the numbers from 1999 to 2007, so we can't blame it on the financial crisis or the particular problems of today.

Even since the recession ended, we see wages for the typical worker continue to decline. From June 2009 (the official end of the recession) to June 2011, inflation-adjusted median household income fell 6.7 percent, more rapidly than it fell during the recession itself (3.2 percent). Median income in 2011 was more than 8 percent lower than in 2007 and indeed median household income peaked in 1999. There's room to dispute how exact these numbers are, but they do show longer-term structural forces at work. A lot of jobs aren't worth as much as before, and they are not being replaced by a comparable number of high-earning slots.

It's pretty common to see new jobs at companies such as General Electric or Caterpillar, and the new jobs cover pretty much the same tasks as the old jobs. Yet the new workers are earning ten or fifteen dollars less an hour, and the companies are putting up with the morale problems involved with having two different tiers of workers.

The numbers also show that earnings from labor are a falling share of total output. In 1990, 63 percent of American national income took the form of payments for labor, but by the middle of 2011 it had fallen to 58 percent. Most developed countries—including Germany, France, and Japan—have seen similar trends. These trends start about 1980, but note also that the income share *for skilled sectors* has been on the rise, going up by about five percentage points in countries where English is the dominant language.

Here is what the trend for the United States looks like:

LABOR INCOME AS A SHARE OF TOTAL INCOME

Note: Shaded bars indicate recessions.

Sources: Bureau of Labor Statistics, National Income and Product Accounts.

© Federal Reserve Bank of Cleveland (Redrawn by Daniel Lagin)

If there is one picture that sums up the dilemma of our contemporary economy, it is that one.

Of course, there is a lot going on within the category of labor earnings. The longer-term trend is fewer jobs in middle-skill, white-collar clerical, administrative, and sales occupations. Demand is rising for low-pay, low-skill jobs, *and* it is rising for high-pay, high-skill jobs, including tech and managerial jobs, but pay is not rising for the jobs in between.

This is not just a story about America, as broadly similar patterns are occurring in the major industrialized nations of Europe. In sixteen major European nations, from 1993 to 2006, middle-wage occupations declined as a share of employment. In thirteen of those sixteen countries, high-wage occupations increased as a share of employment. In all sixteen countries, low-wage occupations rose as a share of employment, relative to middle-wage occupations. There are more really good jobs but more really bad jobs involving exhausting low-wage work too—the kind that might even shorten your life.

The pattern for the really good jobs and the higher earnings is pronounced. If we look at the last ten years and divide the population up by education, there is only one group that has come out ahead with higher wages: individuals with advanced postgraduate degrees. If you had a PhD, on average your earnings went up a little over 5 percent, and if you had an MD, JD, or MBA, average earnings went up a little less than 5 percent. Even for holders of master's degrees, earnings were down an average of about 7 percent. Average earnings were down about 8 percent for individuals with a bachelor's and down 10 percent for individuals with some amount of college but no four-year degree.

It's clear: The world is demanding more in the way of credentials, more in the way of ability, and it is passing along most of the higher rewards to a relatively small cognitive elite. After all, the first two categories of earnings winners—namely those with advanced degrees—account for only about 3 percent of the US population.

Careers

So a smart young person gets a good education and is deciding what to do with it. Why are so many of these people going into finance, law, and consulting?

There is a common impression—by no means illusory—that smart young people from top schools can walk into high-paying jobs in these areas with relative ease, even if they don't have much or indeed any real-world experience. They start at salaries above the US median household income, and very quickly many of them are earning above six figures. In finance they may be paid bonuses of millions within years, at least if they come along during the right years. Beneath all the chatter is a sense that these salaries are possibly unmerited or unjust, because, to repeat an expression I used to hear from my father (he was a businessman of the old school): "I wouldn't trust him in charge of a candy store." If you took a few of these same young workers out of the consulting firm and put them on a factory floor, they probably would be lost. They do seem to be an impractical bunch.

There are some particular reasons why employment opportunities are growing in finance, law, and consulting. Today, laws are more numerous and more complicated than in my father's day, and that increases the demand for lawyers, at least at the top end of the market. A global economy means longer supply chains, and consultants can help businesses track and evaluate those complex operations. Finance is growing in part because the promise of bailouts encourages banks to become larger and also take on more risk. But

I wish to put those (valid) points aside and focus on some more general reasons why smart but underspecialized young people are finding so many good opportunities in these sectors.

As a general rule, the age structure of achievement is being ratcheted upward due to specialization and the growth of knowledge. Mathematicians used to prove theorems at age twenty, but now it happens at age thirty because there is so much more to learn along the way. If you are a talented twenty-two-year-old, just out of Harvard, you probably cannot walk into a furniture factory and quickly design a better machine. Young people *have* made fundamental contributions in some of the internet and social networking sectors, precisely because of the immaturity of those sectors. Mark Zuckerberg needed a good grasp of Myspace, but he didn't have to master decades of previous efforts on online social networks. He was close to starting from scratch. In those cases, young people tend to dominate the sector, but of course that won't cover the furniture factory.

Now take a typical young person, not furniture-machine savvy but just out of Harvard with, say, a degree in economics. She and her parents expect her to earn a high income—now—and to affiliate with other smart, highly educated people, maybe even to marry one of them someday. It won't suffice to move to Dayton and spend four years studying how coffee tables are built.

These freshly minted students will seek out jobs that reward a high "g factor," or high general intelligence. That means finance, law, and consulting. The students are productive fairly quickly, they make good contacts with other smart people, and they can demonstrate that they are smart, for future employment prospects. Working to exercise and demonstrate their general intelligence is in fact the main thing they are good for, and moving beyond this can take quite a few years.

Consider consulting. Take a smart but inexperienced and

underspecialized twenty-two-year-old and ask, "Can you draw up an effective PowerPoint for me?" or "Can you research this new accounting practice or new congressional law?" or even "What is wrong with this business plan?" You might get some pretty good answers. And so there is a demand, from the side of the company, to hire that general intelligence. In due time, the hired smart young people may become leaders or partners in the firm, and along the way they will acquire experience in applying their general intellectual abilities to concrete business problems.

We tend to glamorize these well-paying jobs. If we can set aside the glamour and perhaps our envy, we might notice that our society does not know what else to do with these people, who are otherwise not always very productive.

Fortunately (for them), they really are needed. The more the rest of the world specializes in production, the more that general intelligence can produce some value. Forget about a detailed knowledge of the factory floor, the specialists in Dayton are missing out on the big picture. Try some simple questions and admonitions: "Hey you, think about what you are doing! Are you sure? How about *this*? Are you *sure* that's the best way to treat your workers?" "Do you really understand what is going on in global markets? Come take a look at my PowerPoint."

It often sounds like meaningless or bogus clichés to outsiders, but very often the people in the field do not get it or do not think very conceptually about their own operations. It's not in their training, and in the meantime they have become hyperspecialized in some very particular daily routines, such as mastering how a factory for producing furniture should be run. Every now and then these questions, rooted in general intelligence, pay off and generate a high expected return. The ever so popular management books, which can seem so banal to outside observers, are also attempting to supply critical outside general intelligence. It's a hard set of conceptual skills to communicate and then turn into practice, and thus the demand for consultants—including young consultants—won't be disappearing anytime soon. The flow of business and management books will probably never end.

Labor markets are tough, and not always fair, but intelligence will be rewarded for a long time to come. So will the right skills in STEM fields, finance, management, and marketing, all of which meld together the strengths of diverse intelligences, whether those intelligences are human or not.

3

Why Are So Many People Out of Work?

We may have accepted that machines won't put everyone out of work, and that the rise of intelligent machines will benefit a lot of us greatly, but there can be little doubt that they will also put a few percent of us out of work for some time to come.

Consider what economists call the "labor force participation rate." It refers to the percentage of the civilian non-institutional population sixteen and over who are either working or looking for work.

You can see clearly in the chart on the following page that it has been going down for some time. Human labor suddenly doesn't appear so indispensable, does it? Labor force participation depends on numerous factors, including the business cycle, savings, availability of benefits, and lifecycle and gender considerations. But let's focus on how intelligent machines are likely to make a big difference across the next few decades. And to do that, let's look at the development of one small subsector in the age of the computer.

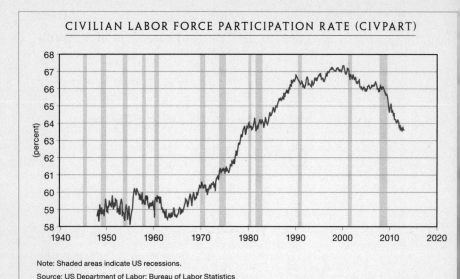

CIVILIAN LABOR FORCE PARTICIPATION RATE (CIVPART)

Note: Shaded areas indicate US recessions.

Source: US Department of Labor: Bureau of Labor Statistics

© Federal Reserve Bank of Saint Louis (Redrawn by Daniel Lagin)

I remember the very early days of computer chess, when I was a kid playing in tournaments in the 1970s. There was an ungainly mechanized beast called Belle, which was occasionally lugged around to tournaments and looked like an oversized adding machine. If the pairings suggested you were to play the machine in the next round, you had the right to refuse, because after all, chess tournaments were for playing human beings. But most decent players were willing to take on the machine, because they could usually beat it and maybe because they were curious too. Belle required a porter to lug it around on a table, someone to turn it on and off, and someone to type in the moves of the game. Every now and then the machine experienced technical difficulties and would spit out nonsense or stop running altogether. Someone had to try to fix it; otherwise the machine forfeited the game. The machine couldn't add much to the game of a good player.

By the 1990s, IBM was working to build a machine that could beat the human world champion. The company hired a lot of programmers, hardware specialists, and of course a few top chess players, such as grandmaster Joel Benjamin. They were "inventing the wheel," so to speak. The Holy Grail was an autonomous machine that could defeat any human. But then, almost without anyone noticing, the nature of the contest changed—the contest for top quality, that is.

By the late 1990s, there were collaborative efforts between computer programs and top grandmasters—the human competitor would consult the program midgame. So was born "Freestyle chess." A top-level collaborative man–machine Freestyle competition meant that a top grandmaster sat down with a computer and the grandmaster thought through the strategy of the game long and hard. The programs still had significant strategic gaps in their play, so a grandmaster supplemented or guided the strategic thinking of the machine but would rely on the machine for accurate short-run tactical calculation.

As the programs improved, Freestyle chess circa 2004–2007 favored players who understood very well how the computer programs worked. These individuals did not have to be great chess players and very often they were not, although they were very swift at processing information and figuring out which lines of chess play required a deeper look with the most powerful programs.

Today, the top Freestyle players fear that the next or maybe even the current generation of programs (e.g., Rybka Cluster) will beat or hold even with the top Freestyle teams. The programs, playing alone without guidance, may not be so easy for the human to improve upon. If the program's play is close enough to perfection, what room is there for the human partners to add wisdom? To improve production, perhaps?

This step-by-step evolution is how intelligent technology will change a lot of industries. At first the machine hardly adds anything and it's really just an investment in building a better machine. At the second step, experts—in the field of the program's operation—will be required to work with the machines, to fill the gaps in what the machines can do. As the programs improve, the next and third step is that the humans understand the programs very well, with a minimum of expertise—but expertise nonetheless—in the relevant industry. These workers will essentially be information processors, albeit with an understanding of context. The fourth and final step is that the human isn't needed much at all because the program on its own is so strong. When that happens, the previous collaborators will look for a new job somewhere else, maybe in an industry that has not yet made such good use of intelligent-machine analysis.

This is a very different picture from the dramatic claim that the machines will replace us all. In most parts of the contemporary economy, artificial intelligence and related concepts remain immature and are not close to dispensing with the humans. Furthermore, the advances have, for the most part, not come from making the machines more like human brains; rather the machines remain complements to humans.

Computer programs do especially well in chess because it is a totally regularized environment where the right answer can be ascertained, at least in principle, by pure calculation. And a lot of those calculations are within the range of current hardware, even on a PC or an iPhone. Chess is relatively easy in comparison to enterprises that need to judge the business cycle or evaluate a person's character. Chess programs can try to find "the best move" without having to calculate the psychology or likely response of the opponent; in this regard chess is unlike the more strategic and psychological game of poker. In poker, the very best players are still humans, because the computers don't know how to psych out the opponent, bluff, or read the "tells" from the guy sitting across the table. The more that an endeavor requires inferences about the mind-states of others, the more that intelligent machines will require human aid. We humans do have our talents.

Intelligent machines aren't going to take over the entire economy all at once, but they will slowly revolutionize our economy. As each economic sector utilizes the new intelligent technologies, the notion of an effective man–machine team will radically change and become extremely diverse.

Many of these man–machine jobs won't be much harder than, in today's world, running a tollbooth on the New Jersey Garden State Parkway, a job performed by both man and machines. Even when the booth is run by a machine alone, every now and then a person has to come over if there has been some kind of misunderstanding, if the driver operates the machine incorrectly, if the driver has run out of money, or perhaps if the machine is broken.

But some jobs will take longer to incorporate the new intelligent technology. For instance, automated disc jockeys are not that impressive. "Denise" has been a disc jockey in San Antonio. She costs $200, plus some programming time. She can tell jokes, report the weather, and look things up on the internet. Yet Denise still requires a script from a human—yes, she is just a program—which includes being told what to look up on the internet. Circa 2012, these disc jockeys are hardly the rage and Denise is described as sounding "a bit robotic." If the idea takes off, disc jockeys like Denise will require a lot of human aid along the way, yet they probably won't need technical geniuses or exceptional performers. This will mean jobs, though they won't be DJ jobs per se, because the program is taking over some of the traditional DJ functions. They will be jobs that will require some understanding of the DJ function and some understanding of what the computer can and cannot do on the radio. They will be jobs that non-geniuses and indeed non-programmers can strive for, at least for a good while.

High unemployment remains a feature of the American economy; in 2011 it was still over 9 percent following the economic crisis, and it dipped just below 8 percent in 2012, but the news isn't as good as that apparent trend may sound. The measured unemployment rate is taken from the group of people who are actively looking for work. About half of the drop in unemployment from 2011 to 2012 came from the fact that many Americans stopped looking for work altogether. If we adjust for this quirk in the definition, the "true" rate of current unemployment is arguably still above 10 percent.

Measured unemployment may well continue its decline, but it has been remarkably slow to fall, and many people remain out of the workforce. Less than a decade ago the picture was very different, with an unemployment rate of 4 percent. At best, we are creating only enough jobs to match population growth in the labor force. Many people with part-time jobs would like to have full-time jobs, and many people are stuck in jobs that give them few chances for advancement. A considerable portion of our labor force is *underemployed*. Economists consider this very poor labor-market performance, and generally don't expect to see full employment returning anytime soon. The crisis of 2008 and the ensuing recession are over, but the deeper structural problems are now showing through. There were short-run job market problems from the fallout from the financial crisis, but it's not just a short-run problem. Our population is growing, but the number of people working continues to fall. This trend was underway well before the recession.

Just how rewarding is work these days? The single best number to look at is the labor force participation rate, which circa 2012 showed that around 63 percent of the labor force was looking for work. Yet not all of these individuals have jobs, so the percentage of individuals in the labor force with jobs stands at around 58 percent. That figure hasn't been so low since the early 1980s. (In those days the number was so low because fewer women wanted to work or had the opportunity to work.)

Those numbers on labor force participation are telling us that, for whatever reason, over 40 percent of adult, non-senior Americans don't consider it worthwhile to have a job. They can't find a deal that suits them.

What if we look only at men in the age bracket 25–64? In the 1950s and 1960s, about 9 percent of this group was not working. Today the percentage of this group not working exceeds 18 percent. For purposes of comparison, in the severe recession of the early 1980s about 15 percent of men in this age bracket were not working, a better record than today. Adult males are seceding from the workforce—or being kicked out—in frightening numbers. Few of these individuals are wealthy playboys. It is no surprise that popular culture today has this image of the male slacker, a young man who lives at home, plays video games, is indifferent to holding down a job, and maybe doesn't run after young women so hard. Most of the measured declines in employment participation have been coming from younger men, not early retirees.

In fact, a lot of older people have been going back to the labor market, largely to help make ends meet. They are working more, and the individuals who should be investing in their futures—younger men—are working less. People are getting accustomed to an existence where they cannot find satisfying work at a wage they are happy with.

The numbers on disability confirm the grim overall picture for the labor market. Ten years ago, 5 million Americans collected federal disability benefit; now the number is up to 8.2 million, at a direct dollar cost of $115 billion a year—over $1,500 for every American household. Yet the American workplace, as measured by deaths and accidents, has never been safer. The point is not whether to call these people lazy bums or to question in which regards they may be disabled. No matter how you interpret the situation of each individual, as a whole they are also people who have not found the modern world of work to be such an attractive option, compared to the collection of disability benefits. It is no accident that workers are most likely to apply for disability following a job loss.

There are also more than two million Americans in jail, in percentage terms the highest rate in the world. I'm not trying to absolve the true criminals of blame for their misdeeds, but you can add them to the list of people who did not, for whatever reason, find a satisfactory labor market outcome.

These problems with labor have gone beyond the general problems with our economy, so something has gone wrong with work itself. In the last ten years the output of the US economy has shown positive growth, and our population has risen, yet the number of jobs has been falling and, as I've already mentioned, median income has performed dismally. Clearly a major structural shift has been occurring.

Male wages have done starkly worse than median household

income. As we all know, women in American labor markets have become more educated, more ambitious, and they have faced less discrimination, all to the better. Women have made some unique, one-time economic gains through these social advances. But for men, from 1969 to 2009, as measured, *it appears that wages for the typical or median male earner have fallen by about 28 percent.* I've seen attempts to dispute these numbers, but the result remains embarrassing; Brookings Institution researcher Scott Winship, for instance, argues that since 1969 the truth is that male wages have fallen by "only" 9 percent. That's still a dismal record.

Imagine yourself as an economist back in 1969, being asked to predict the course of American male wages over the next forty years or so. You are told that no major asteroid will strike the earth and that there will be no nuclear war. The riots of the 1960s will die out rather than consuming our country in flames. Communism would go away as a major threat and most of the world would reject socialism. Who would have thought that wages for the typical guy were going to fall?

It's a stunning truth.

Most economists, when explaining unemployment, cite the Keynesian, neo-monetarist, and "aggregate demand" stories that there is not enough spending in the economy. In these accounts, after the collapse of the housing bubble, asset prices were lower, indebtedness was higher, and people were more uncertain about the future. Most people cut back on their expenditures, and this led to lower sales and lower output and eventually lower employment. This story makes sense, especially as applied to 2008–2009, but it doesn't explain the longer-run trends. For any given shortfall of spending, an economy can respond either robustly or in a very fragile manner. Our labor market has been fragile in response, and this fragility is tied in to the bigger picture about how labor markets are for many individuals becoming worse. Without jettisoning the short-run spending hypothesis, we need a deeper and more fundamental explanation of why the labor market has responded so poorly to the spending decline. Which other factors, acting in conjunction with the spending decline, determine how many people will be getting—and keeping—jobs?

The Great Recession

Let's consider the big job losses of 2008–2009. As the recession came, suddenly there was less spending and a big liquidity crunch. In 2008–2009, a lot of firms were seriously short of cash. Demand was down, their credit lines were restricted, and ambitious plans were being scaled back. Firms didn't have enough cash to keep everyone on payroll or they wanted a bigger cash cushion because of the riskier economic environment. But they didn't fire indiscriminately. They took some discrete steps to figure out which workers were adding the most value, and once they identified the less productive workers, they let them go. These firings cost them effort and they cost them some morale among their better and still-employed workers, who don't always enjoy such close scrutiny of their work. Still, it was necessary and it was done. Unemployment rose dramatically, most of all during 2009. That was a big one-time event brought on by the collapse of spending, but it was also part of a longer trend from the previous decade, namely looking to restructurings and firings as a means of boosting corporate productivity.

Those laid-off workers have been absorbed into new jobs at a rate much slower than is usual following a recession. They can't get their old jobs back, even though the worst of the crisis is over and corporate profits are back up. Most importantly, the new jobs being created are more likely low wage than mid-wage. In essence, the American economy is learning that—for structural reasons—it can't afford as many mid-wage jobs as it used to. Businesses will make higher profits by slotting those workers elsewhere, but not back in other high- or mid-wage jobs, where they had been before.

The financial crash was a very bad one-time event that revealed, rather suddenly, this more fundamental long-term structural problem, namely that a lot of workers had been overemployed relative to their skills. It was a proverbial "moment of truth," and its repercussions continue to haunt many workers. There's some preliminary research by Nir Jaimovich and Henry Siu, two economists, and it suggests that most labor market polarization is transmitted through the immediate mechanism of recessions, which is when those middle class jobs are disappearing. After the recession is over, the lost middle class jobs do not come back.

The burst of the housing bubble and the financial crisis meant that people lost a lot of wealth, and businesses lost a lot of liquidity and a lot of access to future credit. All of a sudden people had to choose in a tighter and more focused way than in earlier years. Consumers had to choose which products they really wanted to buy, and companies had to choose which kinds of workers they really wanted to hire and which kinds of plans they really wanted to emphasize. No longer could people and institutions do "a bit of everything all at once," so future projections got pushed into the present in a very focused and concentrated way. The longer-term truth was that a lot of mid-wage jobs were going away anyway, and we received a quick and painful lesson that this trend is more fundamental and further reaching than we had thought. The more the economy recovers from the recession, the clearer it becomes that a significant structural change has been underway for a longer period of time and that these structural changes are being baked into the look of the recovery.

If we had greater spending and stronger aggregate demand, as the Keynesians and other economists advocate, more of those individuals would find jobs. But the new jobs would, for the most part, not look like the old jobs, because businesses already have rebuilt under the assumption that those old jobs are not coming back. A lot of the new jobs, especially for the less-educated workers who make up such a large chunk of the unemployed, will be low-pay jobs that rely more on brute force or direct personal services, such as running errands for a company or taking care of an elderly person.

When the economy is booming, brute force labor is very much in demand. When your company is swamped with orders, you need people to pull the product off the shelf, wrap a box around it, run a mail sticker, and carry it down to the mail room. That's not picking cotton, but it's still simple physical labor. A worker who previously was useless in the eyes of the labor market will find a job when a needed simple task comes along, such as rushing packages out the door to impatient customers. It doesn't take too much in the way of skill or morale and you don't have to cooperate with too much precision; you just have to get it done. But right now the economy is not booming and we're waiting for more brute force jobs to come along.

It doesn't matter how flexible the wage is in the more complex, less brute force jobs. A manual worker who just shows up at your door is probably not someone you want to hire unless it is already part of a preexisting business plan with broad buy-in from your enterprise and your creditors. The worker might say, "I'll lower my wage demands by thirty percent!" or, "I'll work for nothing!" It usually won't matter. The sad reality is that many of these workers you don't want at all, even if the business plan involves additional labor. Some workers simply aren't worth the trouble unless the demand for extra labor is truly pressing.

I believe these "zero marginal product" workers account for a small but growing percentage of our workforce. At the very least they make it unlikely that we will return to 4 percent unemployment in the foreseeable future.

Ask just about anyone in a human resources department, "What percent of the labor force do you simply not wish to hire, no matter what, no matter how low the wage?" It's quite a few of the applicants, and I can vouch that I have found the same when working as an employer myself. Or look at how many people are turned away by the US military: Drug use, medications, weight problems, bad credit scores, and criminal records make many people ineligible to sign up. According to Arne Duncan, secretary of education under President Obama, three-quarters of today's youth between the ages of seventeen and twenty-four are unfit to serve for one reason or another. *Three-quarters.* Staggering, no?

There's also some fairly direct numerical evidence that the laid-off workers weren't worth as much as we had thought.

Let's look at the last three quarters of 2009, when the increase in unemployment was the steepest. In those quarters, the statistic known as "average per-labor-hour productivity" showed some funny behavior. It's a measure of the results our nation is producing at work. In the second quarter it went up 8 percent, in the third quarter it went up 7 percent, and in the final quarter it went up 5.3 percent— *up* three quarters in a row. Those numbers measure how much output the economy produces for each hour worked. It is an average, which means that if the lesser workers are fired, the number will go up even when total output is going down. For instance, if one worker produced $20 of value per hour and the other $10 of value, the average is $15; firing the lesser worker will raise the average to $20.

After the first quarter of 2009, per-labor-hour productivity *rose dramatically*. Why did that happen? Was it because we invented workable nanotechnology or some fantastic new and highly productive machine in April of 2009? I don't think so. It's because we laid off a lot of workers who weren't producing enough for their level of pay. Bosses were pulling the less productive workers out of the higher-paying jobs. And afterward they didn't want most of them back. That caused average productivity to rise.

Had employers fired a more typical batch of workers, the statistic for average per-labor-hour productivity would not have budged much, even though total output would have fallen due to the lower number of total workers. Recessions were more like that before the 1990s, but since then recessions have been carriers and intensifiers of longer-term labor market trends. When downturns come these days, employers make the tough but correct choice that the future of their companies means fewer of those mid-level, mid-wage jobs. That is one harbinger of our information revolution and how it is reshaping the entire world of work.

There is additional evidence that the currently unemployed have especially low job-finding rates. There is an "inflow" rate to unemployment (those who lose their jobs) and an "outflow" rate (those who get jobs). In the recent recession, it is the outflow rate that has gone especially poorly, because some people seem stuck in a jobless state. That has meant a new and large class of the long-term unemployed. Unemployment is much more concentrated among a group of regular losers—many of whom may now be permanently unemployable—than in previous downturns.

So what happens to laid-off workers, at least those who are still capable of working and willing to work? Whether we like it or not, many of them need to find lower-paying jobs. There are plenty of lower-paying jobs in the world, more than ever before, but here are the rather significant catches:

1. A lot of those jobs are being created overseas. If the job does not require high and complex capital investment, the advantage to keeping that job in the United States is lower.
2. A lot of Americans are not ready to take such jobs, either financially or psychologically. They have been conditioned to expect "jobs in the middle," precisely the area that is falling away.
3. Through law and regulation, the United States is increasing the cost of hiring, whether it be mandated health benefits, risk of lawsuits, or higher minimum wages.

It is hard to escape the conclusion that unemployed young workers will only slowly be reemployed. And the jobs they get will often have considerably lower wages.

Freelancers in Generation Limbo

That slow process of reemployment does not necessarily run smoothly. Two economists, Alan Krueger (as of 2013, President Obama's chief economic adviser) and Andreas Mueller, surveyed over six thousand unemployed workers. They found that many of these workers are not willing to take jobs for much less than their previous salaries. Furthermore, this stubbornness does not much disappear over time, at least not since 2011–2012. However unrealistically, most of these individuals are holding out for a better offer than what the American economy is serving up.

The costs of hiring someone today in the United States are clearly rising, in significant part because of healthcare costs. Health insurance becomes more expensive each year and the Obama healthcare plan makes it more expensive for employers to hire workers without health insurance. A company, at least above a certain size, is supposed to either offer insurance or pay a fine. The mandate in the new law also stops the evolution of cheaper, lower-coverage healthcare plans. You may or may not like the new law as healthcare policy, but it increases the cost of employment and that means fewer good jobs. Kaiser Family Foundation estimates that a health insurance premium today for a family of four averages over $15,000 and within ten years' time could be $32,000 a year or more. That's more than a lot of workers are worth. Keep in mind that the 2010 median wage in the United States for an individual (not a household) was about $26,363. If we force or nudge employers to provide health insurance, a lot of workers won't be worth hiring at good wages.

There are other factors, of course. The federal minimum wage was increased over the course of 2008–2009, with the exact figure varying by state. The ability of disgruntled workers to file and win lawsuits, or even to threaten lawsuits for the sake of being paid off, shows no sign of abating. These may be good or bad things for society as a whole, but without question, we are moving further from a world where it is quick and easy to hire workers. That makes it harder to hire workers at all.

As we should expect from these developments, many American workers are turning to self-employment. For instance, in 2010 an average of 565,000 Americans a month started businesses, the highest rate of the last decade. We're not suddenly more energetic; rather, a lot of these people had a hard time finding remunerative work elsewhere. Starting your own business may seem like praiseworthy creative entrepreneurship, but often it is a sign that labor markets are not absorbing everyone at a reasonable wage.

Rona Economou, age thirty-three at the time, is a typical story. She was laid off her well-paying job at a large Manhattan law firm and the market for lawyers proved tough to crack. After some soul-searching, she responded by opening Boubouki, a small Greek food stall on the Lower East Side, at the Essex Street Market. These days she wakes up at 5:30 A.M., lifts a lot of heavy bags, and can't afford to miss a day of work. It's not clear her project will succeed financially, much less bring her riches, and it also doesn't seem that her life is freer. A lot of future jobs will look like this—that is, they will look more like the jobs we already see in great numbers in developing countries. Sometimes economists glorify the dynamism of "petty entrepreneurship" in developing nations, but in reality most of those people would be better off with steady factory jobs, if only such jobs were to be had.

The US economy has seen a freelancing explosion, including contractors, sole proprietors, consultants, temps, and the self-employed. The government does not accurately count the freelance job market, but anecdotal evidence suggests it has exploded through online postings and temporary employment. It was already the case in 2005, the last time a formal count was taken, that one-third of the US labor force participated in freelance markets in some way. Yet this isn't a utopia of wealthy independent entrepreneurs; in recent times the growth has come in part-time rather than full-time free-lancing. Most of the time it's about making ends meet, not heroically achieving fame and fortune. Over time we can expect these categories to blur, and freelancing jobs will become increasingly respectable and indeed normal, if only because they offer a bit of pay and a bit of personal freedom too. More workers will think of themselves as free agents, and more employers will be keener to make hires without traditional benefits packages being attached to the job offers.

Here are some of the new freelance jobs, or maybe I should call them micro-jobs. Not surprisingly, they are service jobs. The Christenson family hired a worker—a scientist in fact—to come by and drain the "worm juice" from their compost bin. They found him online. Erika Dumaine, then age twenty-four, bought a pair of shoes from Nordstrom and for seventeen dollars delivered them to the home of the woman who wanted them. John Burks of Chicago accidentally dropped his keys down the sewer and placed an urgent online ad for someone to figure out how to fish them out. Within an hour someone was doing the work for eighty dollars.

Among the young there is a growing tendency to postpone adulthood, in part because lucrative job opportunities do not beckon. The new crowd of youngsters is sometimes called "Generation Limbo." They end up living at home for longer, they take freelance and part-time service work—such as in bars or bookstores—or they write part-time for websites. It is less likely that their first or even second jobs will count as potential "careers." I do not presume the limbo generation consists entirely or even mostly of unhappy individuals. They have freedoms and flexibilities that older generations might have envied, and they have the chance to spend lots of time with friends and family. Sex and parties and good ethnic food seem to be everywhere, if Facebook is any kind of guide. Still, the longer-run job prospects for many of this crop of twentysomethings may not turn out to be so great. When we consider how the current generation will do in coming decades, it seems the American economy is not replenishing its "seed capital."

These are just anecdotes, but as I've discussed in the first chapter, the general trend is in the data: Real earnings for graduates (college degree only) are down since 2000.

Today, many of these young earners are threshold earners, meaning earners who are content just to get by and who do not push ambitiously for a higher wage or stronger credentials at every step. Williamsburg, Brooklyn, is full of young threshold earners, although rising rents are starting to push them out into other parts of the city, such as the further reaches of Brooklyn or the Bronx. In Berlin, a city that has become renowned for its supply of threshold earners, it is commonly recognized that a lot of the young denizens simply aren't striving after very much, at least not in terms of commercial job opportunities. A fifth of the population lives off social transfers, unemployment is double the national rate, and, as one commentator suggested, "aspiration can be a negative word." In a wealthy society, sometimes it's just enough to get by and have a good time. It may not sound adventurous or even very American, but we're going to be seeing more of that in the years to come.

Overall, these job market trends are bringing higher pay for bosses, more focus on morale in the workplace, greater demands for conscientious and obedient workers, greater inequality at the top, big gains for the cognitive elite, a lot of freelancing in the services sector, and some tough scrambles for workers without a lot of skills. Those are essential characteristics of the coming American labor markets, the new world of work.

PART II

What Games Are Teaching Us

4

New Work, Old Game

Gaming is a major commercial success story in the digital economy, from *Angry Birds* and *Tetris* to *Grand Theft Auto* and *EverQuest*. It's now a bigger part of our cultural economy than Hollywood movies, even though it doesn't receive comparable attention from cultural commentators—or from economists, for that matter. These games are sometimes thought of as child's play or as a wasteful distraction, but to ignore such developments is a mistake. Games reflect and bring together modern trends in cognition, entertainment, education, and the rapid processing of information. Whether we are talking about entertainment or information, today's world has more and more to do with games. These games are changing the way we interact with one another and how we spend our time. What they teach us will show up in our work lives, as indeed it already has.

Even those on the low end of the income scale can now afford astonishingly complex games that require computational muscle far beyond what was conceivable a decade or two ago. Remember when we wondered if a computer could ever beat a chess grandmaster? Today a good laptop-based computer chess program, available for forty dollars plus shipping—or maybe even for free off the web—is decisively better than any human. Rybka, Fritz, Junior, Houdini, Stockfish, and Komodo are some of the best programs.

Chess is a particularly useful example. Although it is an ancient game, it has become a thriving computer-game category after decades of extensive technical research and development, serving young and old, beginners and grandmasters (and beating them too). This success has made computer chess a rich source of data about human decision making.

We have been learning from games for thousands of years. They have made us better at staying fit. Teachers have used them to help us remember arithmetic, grammar, and chemistry, and now they are used, through apps, to help us lose weight. Games from Twister to the Olympics taught us how to get along better with one another. Now they are teaching us something very new about our limitations and about our intuition. It is a critical development for the future of our labor markets and global economy. The way humans are playing chess with computers now is, I propose, a model that high earners will be emulating in years and decades to come.

To understand intelligent machines and their future influence, we would do well to note Alexander Kronrod's idea that "chess is the *Drosophila* of artificial intelligence." In other words, looking at chess is one way to make sense of the broader picture, just as the fruit fly (the *Drosophila*) has helped us decipher human genetics.

After World War II, computer science pioneers Alan Turing and Claude Shannon both saw that computers would one day play chess, and wrote seminal articles on how it might happen; Turing was brilliant enough to figure out how computers would play chess even before other scientists had figured out computers. Later, chess was picked out as a test case for the development of computer intelligence and given a big financial boost by the computing giant IBM for publicity reasons. Their Deep Blue machine went on to beat the grandmaster, and perhaps the greatest chess player ever, Garry Kasparov. At that time, in 1997, Kasparov still was probably the better player, and he lost to Deep Blue in large part because he lost his cool, but my how things have changed. The last time a serious public contest was tried, in 2005, Hydra crushed Michael Adams 5.5 to 0.5. A half point indicates a draw, and Adams was lucky to come away with that one draw. In the other games he didn't put up much of a fight, even though he was ranked number seven in the world at the time—among humans.

Chess-playing engines, as they are called, have since proliferated, with small businesses developing and selling their products over the internet for less than the price of a pair of running shoes.

There's a nice thing about chess as a test case for the interaction of technological development and humanity: We can measure results with near perfection. We know who won or lost. There are no shortcuts you can take and no winds that may happen to be at your back. We can precisely measure the relative strength of players and programs. With a bit of a time investment, we can discern whether a given chess move was the right one or not. These features make it easier to figure out some truths about where mechanical intelligence will succeed and where it won't. Chess computer play accurately reveals what even extremely talented and well-trained humans, with all the best intuitions, get wrong. Ultimately it is a window into the strategies that high earners will be using to work with intelligent machines.

Machine Games

There is an important and valuable twist in the technological advance of computer games. Now we can watch computers playing chess against one another. But wait, how do we learn anything from watching two machines play chess against one another? What does that even look like?

On May 23, 2011, it's Stockfish version 2.11 against Spark version 1.0. These two programs are trying to "outthink" each other—it's hard to avoid the anthropomorphizing adjectives. Their expertise exceeds that of any human player.

To my scrawny human mind, the game was continually hanging by a thread, with impossibly complex tactics being traded back and forth based on chains of reasoning many moves deep. By the twenty-fifth move, the game was quite unbalanced and it seemed as if every move was a turning point, one way or another. By the thirtieth move, most of the pieces were far from their traditional squares. It didn't look like the intuition-driven chess played by a human grandmaster.

I had a pretty good idea who was winning, but only because another computer, using a lower-grade version of Stockfish, was "watching" the game online and telling me what to think by serving up numerical evaluations of the evolving positions. Of course, I had my doubts. If a weaker version of Stockfish tells you that the stronger, more powerful version of Stockfish is winning—against a different computer program—should you believe it? Maybe it's a bit like a person repeating to himself how right he is all the time. Still, the two Stockfishes together convinced me that they did indeed have a considerable advantage.

By move fifty, Stockfish, playing White, had a clear lead in material but there was a remaining question as to whether White could break through and settle the game once and for all. The human online commentators started asking whether Wikipedia showed the position to be a forced win for White. Other commentators started calling Stockfish "she" ("he" is more common for a chess-playing computer) and gossiping about whether Stockfish was up to snuff in this kind of endgame. The machine made no comment.

I was ready to bet good money on Stockfish winning, but still I wasn't sure. The evaluation functions of the computers usually keep them out of positions where humans can beat them, but the evaluation functions can't always judge when a longer-term breakthrough is possible or not. The time horizons of these programs stretch for only so many moves. Sometimes the computer's evaluation function predicts a win when there is no way to avoid the draw. That's one of the remaining bugs with these engines. A computer could have a lot of extra pieces, but not understand that some blockaded positions simply cannot be broken, no matter how many moves are tried or in what combination. By move sixty-one I was wondering how Stockfish was going to break through and frankly I didn't see how. I had become a skeptic, despite the insistence of the "mini Stockfish," watching online, that its bigger brother had a crushingly decisive position. Something like fear nagged in my gut. Those two black rooks of Spark could form an effective blockade . . .

At move sixty-two, Spark was convinced of the impending defeat and the computer resigned, as its program instructed. Was it really a won position for White? To find out, I brought the moves home to my laptop computer program—Rybka—and played through it a bit further. It seemed Spark was right. Which meant Stockfish and Stockfish's little brother were right too. And they saw it all far before I did. Meanwhile, back at the tournament, only a second after Spark's defeat, a different program, called Junior, started up against the Naum program for another merry match.

Chess grandmasters have coined a phrase—"that's a computer move"—to describe those ugly, counterintuitive decisions made by computers, the moves that surely appear wrong. Yet the machines that produce those ugly moves beat the grandmasters virtually every time. It is true that as computer chess has developed, we humans have actually gotten a little better ourselves—in large part because we are learning from computers—but not better enough to make another contest against a chess computer close. You might say, when it comes to at least this kind of decision making, we are well below the best in the world.

The contests between the machines are already so scintillating, so deep, and so intricate in their tactics that even the best human players have trouble following what is going on nowadays. The moves of the machines show, regularly, how puny and unreliable our intuitions are, even if we spend our decades studying chess.

It makes you wonder if the same is true about the rest of our lives.

Imagine using machine intelligence to guide our daily decisions. The iPhone program Siri, or some new version of her, tells Mary to dump John because he is a lying bastard. Another program tells you to sell your stocks or your home. This social side of mechanized intelligence will not be for the faint of heart.

In computer vs. computer games, it is common that one active tactical strategy confronts another and the games explode with complications. The games, from a human point of view, often appear hair-raising, complex, and downright scary. Imagine being dropped into the driver's seat of a Formula One racecar in the middle of a race. The stress might kill you before you lose control of the 980-horsepower vehicle traveling at 200 miles an hour. It is not so different for a human chess player being dropped into a computer vs. computer chess game—the position on the board will often seem very risky, imbalanced, and out of control. "No human would play that move" is a common commentary on computer recommendations.

As in chess, we can expect to see dramatic gains in the personal and professional lives of people who interpret machine feedback—of all kinds—quickly. A particular personality trait that doesn't come easily to everyone will be needed in a lot of situations: the ability to handle or maybe just ignore the ongoing appearance of stressful situations. For instance, if you're doing a business negotiation, some clever machine may be telling you to "walk away from the deal" a lot more often than you're used to. In the meantime, while you're waiting for them to call you back with a better offer, you will feel the pressure. Perhaps not everyone will wish to go down this computer-aided route, even if it promises more workplace and even dating success. Not everyone wants to go out on a date and have a buzzing iPhone in their pocket indicating "kiss her now" about an hour before such a move might usually be attempted. "Touch her on the shoulder, dummy" will be ignored by many of us. The gains in these cases will go to the hardy, those who can manage stress and embarrassment, but not necessarily to people who act like robots.

When both sides of a chess tournament use computers, new worlds of productivity become possible. Sometimes the answer will be very simple, such as when the chess program spits out a single winning move. But as the use of intelligent machines spreads, such simple victories will become but one subset of a broader segment of human interactions. When one genius machine goes up against another, with or without human collaboration, we will often see some pretty mind-blowing levels of complexity. Is this already happening with automatic split-second, indeed split-millisecond, trading on Wall Street?

It may get riskier yet, as the computers are programmed to play an active, tactical game. The computer is programmed to play for a win, not a draw. We can imagine competing intelligent-machine companies offering programs that seek out an active advantage in a typical human situation. No one rises to the top of the business world by breaking even on a lot of deals, and no one successfully woos a lot of women, or marries the right one, by acting "just okay" or neutral. People know that they need to take chances in complex situations, and they will buy tactical computer programs that help them do this. We're going to generate a lot of hairy, very complicated personal interactions, driven by real-time data analysis and computer intelligence. We'll use the computers to manage our risk-taking and seek out decisive advantages, just as it's increasingly done on chessboards.

Average is over. Some real-world interactions will become a lot simpler and call for conservatism and simple rule-following behavior, while others will become far more complicated and extreme. The case for keeping it simple is plain: Just do what the machine tells you to. Avoid mistakes, hang on to your job, your relationship, your portfolio, or whatever it is you are trying to preserve. Defer to the authority of the beast with intellectual brute force.

But it is easy to imagine intelligent machines changing the way we interact with one another and making significant parts of our world a lot more unpredictable and a lot more passionate. Perhaps we should prepare for the most intense, the most pleasing, and the most dangerous forms of chaos—financial, emotional, and otherwise.

Seeing computers play one another is a remarkable spectacle, at least for a chess lover like me. But it is not the model I had in mind for high earners of the coming years. The model is humans and machines working together and each making use of previously un-celebrated and underutilized abilities.

Working with Machines

It's a long-standing cliché that people guided by pure rationality will behave calmly and dispassionately, perhaps like Mr. Spock of *Star Trek*. A look at chess tournaments between thoughtful, rational players allowed to consult the best chess computer programs, how-ever, suggests otherwise. Risk-taking in humans, especially when engaged in intense competition and challenged to the limits of their abilities, tends to bring out sweat and emotion. This kind of chess is called Freestyle chess, and when you see the people engaged in it, a rather new, passionate physicality in the future of chess is revealed.

The late Herbert Simon, a Nobel laureate in economics, a major force behind computational models, and the father of modern be-havioral economics, felt strongly about the importance of games.

According to his collaborator Fernand Gobet, Simon would start their weekly meetings by asking, "What new data do we have about chess today?"

The latest data is from Freestyle chess. And it is the model to consider if you wish to be among the high earners of the very near future.

5

Our Freestyle Future

In traditional chess tournaments, great care is taken to make sure competitors cannot consult computers or otherwise engage in cheating. Freestyle chess throws these strictures out the window. You can consult books, work with computers, call your grandmaster friend on the phone, bay at the moon, anything. "If they were assisted by the devil, that would probably be covered by the rules," grandmaster Garry Kasparov joked. It's only the moves that count. To make it interesting, though, the time limit is tight (sixty minutes plus fifteen seconds per extra move is common), so the speed at which humans process the machine advice is of the essence. Given the utility of the computers, there is more information to absorb than anyone can possibly get to. Nevertheless, the final decision of which move to play must be made. And a human must make it.

What does it take to be a grandmaster of Freestyle chess? It is something quite different from what makes for the best-rated players of traditional chess.

A series of Freestyle tournaments was held starting in 2005. In the first tournament, grandmasters played, but the winning trophy was taken by ZackS. In a final round, ZackS defeated Russian grandmaster Vladimir Dobrov and his very well rated (2,600+) colleague, who of course worked together with the programs. Who was ZackS? Two guys from New Hampshire, Steven Cramton and Zackary Stephen, then rated at the relatively low levels of 1,685 and 1,398, respectively. Those ratings would not make them formidable local club players, much less regional champions. But they were the best when it came to aggregating the inputs from different computers. In addition to some formidable hardware, they used the chess software engines Fritz, Shredder, Junior, and Chess Tiger. The ZackS duo operated more like a frantic, octopus-armed techno disc jockey than your typical staid chess player, clutching his hands around his head in tectonic concentration. They understand their programs—and presumably themselves—very, very well.

Anson Williams is another top Freestyle player who doesn't have much of a background in traditional chess. Anson, who lives in London, is a telecommunications engineer and software developer. A slim young man of Afro-Caribbean descent, he loves bowling and Johann Sebastian Bach. Fellow team member Nelson Hernandez describes Anson as laconic, very religious, and dedicated to his craft.

Anson does not have any formal chess rating, but he estimates his chess skill at about 1,700 or 1,800 rating points, or that of a competent local club player. Nonetheless, he has done very well with his two quad-core laptops at the Freestyle level. Anson and his team would crush any grandmaster in a match. Against other teams, during one span of top-level play, Anson's team scored twenty-three wins against only one defeat (and twenty-seven draws) across four Freestyle tournaments and fifty-one games.

Along with Anson and Nelson Hernandez, the team is filled out by Yingheng Chen. In her late twenties, she is a graduate from the London School of Economics, not a traditional chess player at all, and now working in finance. She is Anson's girlfriend and has learned the craft from him.

Nelson Hernandez defended his passion for the game thus:

> This may sound like easy work compared to OTB [over-the-board] chess but it really isn't when you consider that your opponent can do the same things and thus has a formidable array of resources as well. It is also quite a trick to orchestrate all these things in real time so as to play the best possible chess. . . .
>
> My role . . . is rather specialized. During these tournaments I am minimally involved and spectacularly indolent as I watch Anson demolish his opponents. Between tournaments I am very actively involved in his opening preparation. This is paradoxical, actually, because I am not a chess player. I approach the game entirely from an analytic, computer-oriented point of view.

Anson, when playing, is in perpetual motion, rushing back and forth from one machine to another, as Freestyle chess is, according to team member Nelson, "all about processing as much computer information as rapidly as possible."

Vasik Rajlich, the programmer of Rybka, considers the top players to be "genetic freaks," though he stresses that he means this in a positive manner; he is a top Freestyle player himself. He sees speed and the rapid processing of information as central to success in Freestyle. In his view, people either have it or they don't. The very best Freestyle players do not necessarily excel at chess and they pick up their Freestyle skills rather rapidly, sometimes within twenty hours of practice. He refers to Dagh Nielsen, one of the top Freestyle players, as operating in a rapid "swirl" during a Freestyle game.

Some players enter these events using a chess engine only, set on autopilot and not using any additional human aid. These "teams" do not take the top prizes, and they are looked down upon by the more enthusiastic partisans of Freestyle. Dagh Nielsen estimated that the Freestyle teams were at least 300 Elo rating points better than the machines alone (a measurement of players' relative skill levels), although that was a few years ago. Nelson Hernandez estimates a 100–150 rating point advantage, which is like the difference between the number one player in the world and the number seventy-five player.

Without the programs, even the strongest grandmasters would be unlikely to qualify for the Freestyle finals.

Top American grandmaster Hikaru Nakamura was not a huge hit when he tried Freestyle chess, even though he was working with the programs. His problem? Not enough trust in the machines. He once boasted, "I use my brain, because it's better than Rybka on six out of seven days of the week." He was wrong.

Freestyle teams study the opening moves their machine opponents have made in previous games because, as Kasparov has observed, an initial advantage in Freestyle chess usually means an eventual victory. The players also know the weaknesses of particular engines and how one engine can at times offset the weaknesses of another. They understand time management—at which points in the game a program should spend more time thinking—better than do the programs themselves. Most of all, they understand computer search as a process. They have an uncanny feel for when one of the programs is spitting out a result that might "flip" if that position were to be examined with several more strong moves. They play out those moves on another copy of the program, or on another program, and so they hope to lure the other team into a trap. If the other team is blindly following a program's recommendation and is unaware of the possibility of the forthcoming flip in evaluation, the more perceptive team might just win. In other words, the humans are improving upon the program and searching some lines more intensively than the machine would know to do on its own.

Arno Nickel, a Freestyle expert, noted "in order to win you have to create something special." Vasik Rajlich says that, to date, the gap between the programs and the top Freestyle teams has stayed more or less constant. The human element really does add something, at least for the time being, although he too wonders how long this will remain the case.

The top games of Freestyle chess probably are the greatest heights chess has reached, though who actually is to judge? The human–machine pair is better than any human—or any machine—can readily evaluate. No search engine will recognize the paired efforts as being the best available, because the paired strategies are deeper than what the machine alone can properly evaluate.

I've spent many hours playing a form of Freestyle chess at home. The Shredder program on my iPad has a performance rating of about 2,200, which makes it Master strength—that is, pretty good but not comparable to a world-class player. (It can play better than that, but it becomes much slower; besides, I like being able to beat it sometimes.) My procedure is simple. I play Shredder against itself, but every now and then I overrule the decisions of the program. In essence it's "me plus Shredder" against Shredder. The human–computer team usually wins. At four or five crucial points during the game, I override the strategic judgment of the program and come up with a better move, or at least what I think is a better move. Then I let the superior execution of the program take over. This works maybe four times out of five.

For me plus Shredder to beat Shredder, I don't have to be as good as Shredder, I simply have to understand the game well enough. I also have to be "meta-rational," to borrow a term from decision theory. That is, I must realize that in most situations the judgment of Shredder is simply better than my own, and defer accordingly. I am most likely to succeed in overriding the judgment of Shredder in complex strategic positions, in some endgames, when the program is fooling around with questionable opening choices (1.e4 g5?), and when the program is getting greedy for material. The iPad version of Shredder also has a fairly limited time horizon and sometimes I can see further ahead than the machine, though in these cases I must be very careful. I can't outcalculate the machine unless it boils down to the machine's shorter time horizon, and I don't always know if the length of the time horizon is the key issue. Still, gambling on this matter succeeds more often than not, though it is how I end up losing when I do.

This Freestyle model is important because we are going to see more and more examples of it in the world. Don't think of it as an age in which machines are taking over from humanity. After all, the machines embody the principles of man–machine collaboration at their core—even when they are playing alone.

Machines Are Us

To start the game, the computer consults its "opening book." Think of the opening book as a very large database of high-quality games that have already been played and recorded, including (possibly) some variations that perhaps haven't yet been played but are added by the designers of the program to give the computer extra depth. Data storage is cheap, so the opening book of a good program contains many millions of variations. In other words, the computer knows almost everything there is to be known about the standard chess openings. It is a little like embedding a lot of historical conversations into a database to help it pass a Turing test (giving it the ability to mimic what a human conversationalist would sound like).

When the computer plays the opening, it isn't the computer "thinking," but rather it is the computer having recorded the best of human knowledge about chess openings to date. The program is consulting memory and playing without calculating; instead it searches the book for which opening moves have worked well for top players in the past. It's like playing against a human who can (rapidly) consult every book and every chess database on the planet. It depends on the choice of opening, but it is common that for the first twenty moves—thirty moves or more in some variations— you are playing against near perfection when you confront the computer. Nonetheless, it is the near perfection of collective human experience, not the computer's calculation. If you don't know these openings as well as the program does (ha), you will start with a decided disadvantage. The game is already "Fabulous You vs. The Mighty Human Plus Computer."

Some critics of computer chess think that computer reliance on the opening book is wrong or unfair. Indeed, most programs allow you to turn off the book, in which case the computer has to think through the opening moves on its own. Even the very best computers aren't so good at doing that, because in the early stages of the game there are an especially large number of variations, and long-term strategy is usually paramount over tactics. For the first fifteen moves or so of most games, a reasonably strong master (even more, a top grandmaster) is a better player than a top-class program without its opening book. It is a little like trying to, say, write something meaningful about business or government or education without knowing any history.

There's also a lot of hidden cooperation in old-fashioned human vs. human chess matches.

In human vs. human play, the value of pulling a surprise in the game's opening phase has gone up. Let's say I am familiar with your first twenty moves against my prepared variation and I can introduce a novelty on move twenty-one. Furthermore, I've been able to foresee that the game might reach this point. In my free time, before the game starts, I have asked the computer to give me its best recommendations from move twenty-one onward, against all of your possible responses. I come to the game armed with that computer analysis. In essence, for a critical segment of the game, it is you playing against me plus the computer. Compare this to the job candidate who goes into an interview with some profit projections he has pulled from an intelligent machine.

If I can get the game into territory that I have foreseen and you have not, you are playing against "me plus the computer." In my old competitive chess days, a surprise opening innovation I came up with simply meant "you against me and my home preparation," which is a far less formidable challenge for an opponent.

When preparing to play against each other at the highest levels, top grandmasters spend a large chunk of their time searching for these surprise innovations, to shift the balance of power in a game. They assemble a trove of opening innovations and analyze those moves in advance with the programs. They then hope to lure their opponents into those situations, thereby achieving "me plus the computer vs. you" for part of the game, which is a big advantage. In a sense, they are trying to get to that place in the ring that has a special machine weapon that will ensure victory. The better the players, the more important this potential edge becomes.

Regular chess players who have gained the most from chess engines are those who understand how to train with the computer and how to learn from the computer. The advent of the programs rewards players who have very good memories and conscientious study habits; a player who can't remember his or her prepared lines won't benefit from this strategy. A human with a good memory can carry the learning of a computer program very effectively and apply that learning in a spontaneous fashion in, say, a job interview.

Indeed, in plenty of real-world situations the immediate command over factual or analytical material brings a big edge. Discussions in meetings, strategies and reactions during sales calls, lawyers arguing in front of a jury, and managers in volatile, voices-raised personnel situations all try to draw upon preprocessed information at a moment's notice. In all those cases, it matters more and more what workers have learned from the computer, or not, and how well they remember computer-derived information and advice.

Freestyle Masters

In Freestyle chess, this competition to find the opening novelty is explicit and it leads to a very fine division of labor. Let's go back to Nelson Hernandez, one member of the Anson Williams Freestyle team. For seven years, Nelson has been spending about 20–25 hours a week assembling one of the very best openings databases on earth. He finds top grandmaster and computer games, captures them, records the results of the games, and checks to see that the recorded information is not buggy. It has so much data that he doesn't want me to tell the world how much, for competitive reasons, but I can report that when I heard the number I was taken aback. He is completely dedicated to this one function on the Freestyle team. Armed with this database, his team has comprehensive knowledge of which opening moves have worked out well in past top games and which have not. The database allows them to execute some of the very best openings the chess world has seen, and for many moves into the game.

When I met Nelson, I didn't have the right words to praise him: "I can't say that you're one of the greatest chess players ever, but you're something really impressive. I'm just not sure what." He smiled in return.

By the way, the formal name of the Williams-Hernandez-Chen Freestyle team is not quite public information. I know some of the iterations of the team name, but overall they don't want to tell me and they don't want me to tell you. They keep it as less-than-fully-public information, and they change their name with each tournament. That means other Freestyle teams will find it harder to prepare to play against their openings and harder to know whom and what they are facing in a given round.

Secret teams. Board games. Code names. Does this all sound a little too much like child's play? Could the Freestyle chess model really matter all that much? Am I crazy to think direct man–machine cooperation, focused on making very specific evaluations or completing very specific tasks, will revolutionize much of our economy, including many parts of the service sector? Could it really be a matter of life or death?

Medical diagnosis is crying out for further applications of artificial intelligence. Computers already can diagnose some medical conditions, often using a technique known as artificial neural networks, or ANN. The Mayo Clinic has used ANN programs to assess whether patients have endocarditis, a kind of heart infection, and more accurate diagnoses have saved some patients from unnecessary invasive procedures. General Electric and Artificial Intelligence in Medicine, Inc. are developing further software programs for diagnosis.

Pap smears are very useful for catching cervical cancer, and since the 1990s automated imaging systems have been used to screen pap slides. Image-searching software scans slides for signs of abnormal cells far more rapidly than a human could, but it's not all about the machine. The software identifies the images that human experts need to look at more closely, and there is evidence that these man–machine collaborations outperform the humans working alone, whether in terms of accuracy or speed.

Machines can double-check human diagnoses, catch the mistakes of very tired doctors, and keep up with and store new developments in the medical literature. And by the way, that literature doubles in size every few years, much more rapidly than any human can follow. For all of our scientific and medical progress, misdiagnoses remain common.

Of course, it can be irresponsible to rely completely on the computer's pattern recognition skills, since the human eye will pick up image errors or flubbed data inputs in a way that the machine may not. But a man–machine *team* is less vulnerable to machine oversights; human supervision is often stipulated by the companies that market medical software.

One medical innovation would run a patient's reported symptoms through a Watson-like software program and see what might be wrong, drawing upon extensive databases. But can the computer ask follow-up questions to the patient properly or guess where the patient might be lying or exaggerating in the description of symptoms? Can the computer explain the diagnosis properly to patients with a broad variety of backgrounds and educational experiences? Not anytime soon, and so we are back to collaboration.

It is clear that for the collaboration to work, we need to have a very smart machine. But, if the machine is already in place and plugged in, how expert does the human have to be? When the worker has to be a highly paid physician, a collaborative team can be costly, even if it improves health outcomes. The world—not to mention the American Medical Association—is pretty far from accepting this fact, but the person working with the computer doesn't have to be a doctor or even a medical expert. She has to be good at understanding and correcting the computer's mistakes, which is a very different skill. This will involve some knowledge of medicine, brain scans, or whatever, but it is a less comprehensive medical knowledge than what a prestigious MD would have. It may well involve more knowledge of smart machines, how they work, and what their failings are likely to be.

In the meantime, there is plenty of unauthorized or "gray" collaborative activity in medical diagnosis. Who or what is the most frequently consulted "doctor" in the United States today? It is Google, which gives the user access to over three billion medical articles on the web. Do you not feel so great? Many millions of people type their symptoms into Google and see what comes out the other end. Only then do they decide whether to see the doctor, or visit the emergency room, or perhaps to take more of their medication, or stop taking it altogether. The age of the mechanical doctor is here, whether we like it this way or not.

Making Freestyle Work

How good is private use of Google as a diagnostic device? We still don't know. How many home users know they might get a better set of results with a search procedure like " 'Metabolic Syndrome' site:edu"? Probably not so many.

One study by Hangwi Tang and Jennifer Hwee Kwoon Ng looked closely at twenty-six diagnosed cases and entered the symptoms into Google. Google gave the correct diagnosis in 58 percent of these cases, and in some of the other cases it simply wasn't specific enough. The study did not consider the possible costs of incorrect or misleading results, so we're still far from evaluating this rather large-scale experiment in new medical institutions. If nothing else, it's an argument for proceeding with more regularized and authorized forms of the collaborative approach in medicine. We already have computerized doctors, and that illustrates the power of information technology to spread rapidly; the next question is how good and how reliable our mechanical medical servants are going to be.

Our educational standards, and in the longer term our regulatory standards, will need to change. Our systems of certification and credentialism haven't caught up to the present and forthcoming reality of man–machine teams, most of all in medicine. But that is after all just one sector.

The notion of a diagnostic test is quite general. Diagnoses are performed by managers, lawyers, third-grade teachers, and bank loan officers, and each of those jobs offers an opportunity to exploit the potential of machine intelligence.

Two recent comments on my blog, *MarginalRevolution*, summed up the new world of work pretty well. The first is from a blog reader who identifies himself or herself as Wil W:

> I think we underestimate the need for partial IT laborers in the next five years. There is a growing need for employees who are not working in IT that have some IT training and ability. These employees would not be working on IT stuff more than 20% of their time, but that 20% will be spent using and fixing technology to the productivity benefit of the business as a whole. Often this is called the decentralization of IT, but it has more to do with the tools the people are using in the workplace.
>
> Imagine for a second the value of a worker on the assembly line with no computer training or ability. That is what we used to have 99% of the time. Yet computers

have become a greater and greater part of the process. The value to the company of these workers is decreasing or at a minimum not increasing even though their productivity is increasing.

Now imagine a new worker. He does not have a computer science degree, but has training and computer abilities above the average. Most important he has an inclination to be able to handle computers outside of a pre-scripted process. The value of this worker to the company is not only greater than the first worker, it increases as the use of computers increases.

I am not saying that education is the answer, but that IT is and will be more in the future spilling over into more labor centered areas and that those employees with or acquiring these skill sets are going to be paid better (or should be if allowed.)

Here is the second comment from reader JWatts:

I work in the automation controls industry. My specialty is HMI (human machine interface) development. So I've got a lot of experience in exactly what you are referring to. And I'd say the work that the controls industry has been doing in converting over to heavy automation in the US is spilling over to other fields as well. The most obvious examples that most people see are ATM's, self-checkout kiosks at groceries stores and the like, and graphical menu displays at fast food restaurants.

It's easy to think the menu display at a restaurant is trivial, but it is a huge waste reducer and labor saver. It results in significantly fewer incorrect orders and a faster process overall. . . .

A worker with no computer knowledge is of little value in a modern American factory. Even the tow truck drivers have computers on the tow trucks. There are literally more computers than people in almost any high value product factory. (And all of the low value product factories are gone or are on their way out of the US/Canada.) . . .

I doubt "education" is the answer, because it's superfluous. An employee that can use a smart phone effectively can use a modern industrial computer. If you can play any kind of strategic video game well, you are more than competent to run a modern line. The fact is that society is busy actively training up the population in an ongoing fashion. And more formal education is unlikely to be of much benefit.

No operator I've worked with needed to know how to write ladder logic code, though some could and were pretty good at it. If I'm doing my job half as well as I should, the HMI and PLCs (factory computers) should automate the repetitious parts of the process. The operator's most valuable contribution is as a troubleshooter and a well designed system makes that process straightforward. That kind of troubleshooting isn't really something you learn in college. However, you learn it just fine customizing your latest iPhone.

What are the broader lessons about the Freestyle approach to working or playing with intelligent machines? They are pretty similar to the broader lessons about labor markets from chapters two and three:

1. Human–computer teams are the best teams.
2. The person working the smart machine doesn't have to be expert in the task at hand.
3. Below some critical level of skill, adding a man to the machine will make the team less effective than the machine working alone.
4. Knowing one's own limits is more important than it used to be.

We also can use the concept of man–machine collaboration to define the difference between a worthless or "zero marginal product" worker and a potentially valuable worker. The worthless worker is one whose cooperation with the machine makes the final outcome worse rather than better. A potentially valuable worker offers the promise of improving on the machine, taken alone. In the language of economics, we can say that the productive worker and the smart machine are, in today's labor markets, stronger *complements* than before.

The effect of those powerful new complements in our economy constitutes the road up from the great stagnation that I have written about elsewhere. But for all its rewards, it will be a scary path for many of us.

6

Why Intuition Isn't Helping
You Get a Job

L ove and romance is one area where human intuition is highly
imperfect and sometimes even dangerous. Apart from extreme crimes of passion, we take a lot of wrong turns in our
pursuit of a good partner, thanks to following our feelings and im-
pulses. Online dating and its machine intelligence has shifted the
rules, but is it merely replacing one set of romantic mistakes with a
new set of biases and oversights?

The major dating services use artificial intelligence to recommend
romantic matches, and this makes a difference because those selec-
tions very often differ from what individuals would have chosen for
themselves. For instance, eHarmony has gone from analyzing a few
dozen variables per user to six hundred variables per user (a figure
that may well be obsolete by the time you are reading this). Those
variables include not just profile information but also how frequently
users log in, whom they search for, and which profiles they actually
decide to write to. One problem, of course, is that a lot of the data are
based on self-reporting, and the answers of the respondents may be

careless or outright lies based in the hope of being offered better and more exclusive matches. The algorithms also tend to select for similarities, and that may encourage initial meetings, but it is less clear it will produce happier marriages thirty years down the line.

The current case for the software is not that it can turn love into a statistically measurable algorithm. We'd be foolish to ignore the importance of conversational rhythms and sniffs and smells and sexual chemistry and how long-term compatibility evolves over the years and decades. Nonetheless, the software is getting users to look twice at some prospects they might otherwise have passed over or never seen in the first place. Here is one romance from Match.com:

> Viewing the profile, Cambry, who is black, saw a pretty young white woman who lived nearby and seemed to share his interest in music. He sent her a short e-mail to say hello, and within a day received an e-mail from Karrah O'Daniel, an opera singer. Their first date was a flop, but they made it to a second date, and soon Cambry and O'Daniel were getting serious. It turns out they had attended the same music school, but never met there. They took to playing pieces by Franz Liszt together, recording videos that they would post on YouTube. Six months later, he proposed. The two are to wed on October 1 at a church in Minnesota.
>
> As often happens in love, the woman Cambry fell for is not exactly the woman he thought he wanted. "I wasn't expecting that the person I was going to marry would be a white woman from Inver Grove Heights, Minnesota," he says. Nor did Cambry fit neatly into O'Daniel's idea of who she would marry. According to her Match profile, she was looking for a man between the ages of 21 and 26

[Cambry was then 28]. Cambry, by O'Daniel's own standards, was too old for her. In fact, Cambry and O'Daniel never really searched for one another at all. They were introduced by the algorithm.

In chess we know the programs can beat the humans, but their skill at choosing dates is harder to assess, and there are no rigorous results available for study. One happy couple doesn't prove that computers are the best matchmakers, nor does it show that we are better off with matchmakers, be they human or nonhuman.

But it is easy to recognize the value of having a matchmaker nudge the indecisive into, well, growing up. You can't keep shopping forever, though humans have long seemed to want to do so. It is common to continue sampling profiles and dreaming on about a perfect mate but not actually dating anyone. Surely this is not the best use of the technology. Perhaps most importantly now, dating algorithm technology can help us realize the errors in some of our personally generated choices. If human intuitions have been so flawed in the highly intellectualized world of grandmaster chess, what might we expect in the area of passionate romance?

I cannot help but think of my own experience from 2003, which precedes the algorithms. I emailed a woman who listed her profile as politically liberal. I would not describe myself the same way (I am more a mix of libertarian and conservative), and she did not know this about me at first and probably, at the time, would not have liked it. I didn't intend to mislead her; I thought it was rather obvious, but to her it wasn't, not even after reading my vita online. She accepted my offer for a date and we have now been married for over ten years. The medium forced at least one of us—probably both of us—out of our usual intuitions and to our mutual benefit.

Years later, when doing the research for this book, I read that the scientists at Match.com have discovered that a conservative is, on average, more willing to email a profile listed as politically liberal than vice versa. So if a profile listed as conservative has a record of emailing lots of liberals, the service will recommend more liberals to that conservative. The conservative won't know why those profiles are popping up as often as they do, yet this matching technique stands a good chance of working.

Will we be willing to learn how to find love from the machines? One New York woman was firm: "The Match algorithm should have figured out that I don't want a forty-five-year-old from *New Jersey* [emphasis mine]." It remains to be seen how she will respond when a machine as smart as Deep Blue sends her to a forty-two-year-old in Kansas as a potential mate.

In our pursuit of romance and long-term partnership we humans tend to avoid unfamiliar complications. We often celebrate our commonalities and avoid the complications that come from dwelling on our differences. This is one place machine intelligence can help. Machines have no fear of the unfamiliar.

Decision Making

Perhaps the most important result from computer chess is an insight into the depth of cognitive problem solving. The top programs have a quality known as "contempt aversion," which causes them to avoid draws and to seek out unfamiliar complications. More importantly, unlike many humans, the programs have no fear of complications. When two strong programs are playing each other, very often you get just that—complications. Many computer games serve up counterintuitive piece placements and patterns. Sometimes the board will appear in chaos. Why were so many pawns moved forward? Why so many gambits? What is the knight doing on the side of the board at this point? Why are so many pieces left under threat of immediate capture at the same time? Aren't these players out of control?

Are they from New Jersey or something? I just do not do things that way!

Biases, such as toward the familiar, are something we have long tried to overcome. In the growing field of behavioral economics, researchers measure the biases behind individual choices as judged by some external standard. We've learned—or we think we've learned—that individuals overestimate their degree of influence over events, and anchor too much on one piece of information when making decisions, among many other human errors and biases. The last time I looked, the list of cognitive biases on Wikipedia had forty-eight entries.

But even after all this work and all this evidence, nagging questions remain. When it comes to measuring a human bias, are we *sure* that the researcher is correct and that the individual choice is wrong? I have a lot of funny habits that I think serve me pretty well. It probably doesn't make sense to an observing researcher that I leave so many piles of books around the house, but to me it's a pretty efficient way of organizing the information and keeping track of the books. Maybe I'm right, or maybe my wife is right when she says to clean up the piles, but either way maybe we don't know.

We see similar dilemmas in the more systematic literature. Remember that old rule of thumb "A bird in the hand is worth two in the bush"? Economists often consider it a "bias" that we value commodities we already own much more than commodities we might acquire (this bias is also known as the "endowment effect"). But for all its apparent irrationality, maybe this is an inescapable part of anyone's ability to be loyal to friends and family. Maybe part of true loyalty is that we can't always apply it selectively on a moment's notice. In that case, these endowment effects may be an essential part of the good life rather than a signal of our irrationality. I'm not saying we know this for sure, only that the models we economists devise don't really settle it.

When economists investigate human rationality, they are often too dependent on arbitrary stipulations about what is rational and what is not, expressed in the form of models. An economist might write down some mathematical axioms and then find that human behavior falls short of those axioms. But how convincing were those axioms in the first place for complex and multidimensional human problem solving? A lot of the research in this tradition isn't convincing, no matter how brilliant the investigator. Other economists rely on artificially constructed laboratory investigations to try to measure human rationality or lack thereof. They use inexperienced undergraduate subjects, who are not always taking the problem-solving exercises seriously, and the prizes for good performance are relatively small. For obvious reasons, it is harder to run comparable economic experiments with corporate CEOs and the relevant real-world decision makers. Furthermore, there are some papers where the experiments are run with CEOs and often the results are the same as when the experiments are run with the students. But isn't that a problem with the experiments rather than a virtue of the method? We know that the CEOs and students are not interchangeable in the real world, so why should they be interchangeable in the experiment? If they are interchangeable in the experiment, doesn't that mean the experiment isn't capturing the relevant real-world context? No matter what you think of these experiments, we would like to supplement them with data from real-world situations with real prizes on the line.

The nice thing about computer chess is that we have clear-cut standards, albeit not perfect standards, for good chess moves and bad chess moves. So, putting human play through the lens of the engines, what do we learn about human intuition?

Let's look to some creative researchers, first Ken Regan. Professor Regan and I were boyhood friends, playing chess together a few times on the same team in the US team championship, in Atlantic City, New Jersey. Until recently, I had not seen Ken since I was fifteen years old. Of all the American chess prodigies who followed in the wake of Bobby Fischer, Ken was arguably the most creative. Ken would play the kind of unusual moves that didn't occur to other players, weaving complicated and beautiful positions, validating sequences that other players didn't think were worth considering. Everyone knew that if they looked at a Ken Regan game, they would see something interesting and perhaps come away with a headache as well. Yet Ken quit professional chess at age twenty-two and turned his attention to math, receiving a PhD in mathematics from Oxford University. He is now a professor of computer science at SUNY Buffalo, and he has spent a good chunk of his professional life working on the famous "P versus NP" problem in mathematics, one of the most important unsolved puzzles in computation (in short, the problem asks whether finding a solution from scratch is in principle harder than verifying a potential solution in hand).

In 2006, Ken turned his attention back to chess, motivated by the growing quality of the programs and captivated by the data they had the potential to generate. Since 2006, he has been constructing a database of games from chess history, focusing on the top players but covering many lower-rated players too. Ken takes each game, captures it in his program, and asks Rybka to analyze the game and to assess the good and bad moves of each player. This is a remarkable experiment in human intelligence. The quantity of data on people's choices makes this an exceptionally valuable window into intellectual performance, into what we humans get right and what we get wrong.

Unlike in a lot of economic models and measurements, Ken has Rybka, a reliable metric for judging human decision making in chess. At any point in the game, Rybka recommends a favored move, and it can calculate how much an alternative move lowers the player's absolute position quality in the game. For every player, any particular game, or any particular tournament, Ken can compute what percentage of that player's moves correspond to Rybka's recommendations and the size of that player's average error, again relative to Rybka's recommendations.

The prizes at chess tournaments are often substantial, and tournament success determines access to future prize money and other career advancement opportunities. Pride, reputation, and rating points are further incentives for good performance, and these chess players are experienced at the task at hand. These are peak performances of the human intellect.

Ken has yet to publish his results, in part because he is still capturing games and compiling additional data. Nonetheless, some patterns have come in through his investigation. I visited Ken in Buffalo, he showed me how he runs his tests on the computer, and he explained some of the results.

Most of all, Ken is impressed by the overall reliability of human decision making. Rybka is of course better at chess than humans, but both the good moves and the mistakes of the humans fit regular, understandable patterns. There is a sense of rationality and order to human error. These same human beings are making decisions about love, business, and which new car to buy, and there is something comforting about what shows up in these patterns.

For instance, players are at their toughest when it matters most. A player is least likely to make a major error when the game is tight, and if anything, players do their absolute best when they are faced with a slight disadvantage in their position. When players are decisively up or down, they don't seem to think or concentrate with the same facility. Again, this is a sign of human rationality, at least if there is some need for a conservation of effort.

Before these investigations, Ken expected to find evidence for a Nassim Taleb "Black Swan" model of cognitive failure. That is, a lot of errors coming out of the blue. But in fact, radically surprising "Black Swan" errors don't play much of a role in the final outcome. Most games are decided on the basis of the accumulation of advantages, and the level of error is fairly well predicted by the relative skills of the players. Ken finds these results all the way down to the level of a 1,600-rated player, which would be a middling club player in most cities (he has yet to look at the games of worse players).

Players also have consistent styles. For instance, Vladimir Kramnik, world champion from 2000 to 2007, plays with an especially high level of accuracy relative to what the computer suggests. His expected error is extremely low—for a human being, that is. Yet Kramnik also does not pose his opponents many problems over the board. He does not cause an inordinate number of mistakes, and he does not crush his opponents' egos or cause them to choke up and collapse in the face of mind-numbingly complex positions. It's not that hard to play fairly well against Kramnik. He creates calm positions with a lot of good, long-term strategic moves.

Under Ken's measure of the most "nettlesome" player—the

player whose play most pushes his opponents into an above-average error rate—prodigy Magnus Carlsen scores the highest. If you play against Carlsen, you're much more likely to make major blunders, as the game more rapidly becomes a minefield of complications.

Ken may seem like just a normal nerdy computer science professor but he, like some of the Freestyle chess players, is marking a new age to come for intelligent machines. In addition to his concrete results, Ken is doing some work of far-reaching social importance. Ken's research reveals the extreme accuracy with which intelligent machines are able to evaluate human performance. Imagine if we had a comparable system to measure the quality of doctors.

Vasik Rajlich, the programmer behind Rybka, gives a more pessimistic spin to what we have learned from the chess-playing programs. In Rajlich's view, the striking fact about chess is how hard it is for humans to play it well. The output from the programs shows that we are making mistakes on a very large number of moves. Ken's measures show that even top grandmasters, except at the very peaks of their performances, are fortunate to match Rybka's recommendations 55 percent of the time. When I compare a grandmaster game to an ongoing Rybka evaluation, what I see is an initial position of value being squandered away by blunders—if only small ones—again and again and again. It's a bit depressing.

Rajlich stresses that humans blunder constantly, that it is hard to be objective, hard to keep concentrating, and hard to calculate a large number of variations with exactness. He is not talking here about the club patzer but rather the top grandmasters: "I am surprised how far they are from perfection." In earlier times these grandmasters had a kind of aura about them among the chess-viewing public, but in the days of the programs the top grandmasters now command less respect.

When a world-class player plays a club expert, the world-class player looks brilliant and invincible at the board. Indeed, the world-class player does choose a lot of very good moves. At some point his superior position starts "playing itself," to use an expression from the chess world, and just about everything falls into place. When the same world-class player takes on Shredder, to select an appropriately named program, he seems more like a hapless fool who must exert great care to keep the situation under control at all. And yet it is the very same player. That gap—between our perception of superior human intellect and its actual reality—is the sobering lesson of the programs.

So what? Haven't thousands of articles from psychology and behavioral economics outlined major weaknesses in human perception and decision-making abilities? There are the works of Daniel Kahneman, Dan Ariely, and many others. Haven't we all heard about "nudge," the concept so eloquently outlined by Cass Sunstein and Richard Thaler? In that worldview, experts know the biases of other decision makers and design the choice architecture to manipulate better human choices, such as changing the default options for which pension plan you will enroll in.

Yes, but the chess result differs. Computer chess is pointing out some imperfections in the world's experts, or you might say it is pointing out imperfections in those who, in other contexts, might be nudgers themselves.

Human intuitions can misfire even when we study the world's best players, who've each been trained for decades to think rationally and compete for high stakes while drawing upon centuries of human experience with the game. The biggest problem is not outright gross blunders, but rather humans spending too much time thinking about the moves that "look good." It is precisely our reasoned, considered judgments that we should be more suspicious of.

Signs of Improvement

There is one other bias in chess that we are now able to measure. Researchers from Sweden, including Patrik Gränsmark, Christer Gerdes, and Anna Dreber, using data from actual chess games, have shown that male players are more impatient and female players are more likely to use up too much of their time thinking. The researchers also find that female players are more risk-averse than are male players; this is consistent with studies of investment behavior and portfolios from the financial world. They have shown that to get male chess players to take even greater risks, you simply have to pair them against an attractive female opponent. The risk-taking and aggressiveness of the male player goes up, though not in a manner that increases his chances of winning the game.

Not everything that is true in the chess world is true in all broader spheres as well, but such results point toward the future of cognitive research. We can get more reliable data from more places. We have intelligent machines that can help us collect and analyze it. Research on chess play is an insight into what machine intelligence is contributing to the forefront of scientific research into human nature—and to our gross national product.

Despite all our bad moves, from chess to the game of love, the good news is that we can learn and reverse some of our excess reliance on intuition.

Ken Regan's results show that the best human chess players are consistently getting better with time. Before 1970, chess ratings were not calculated, but Ken constructs a data series by comparing the play of earlier chess masters to Rybka's move-by-move recommendations.

If we consider the best player from the 1850s, the American Paul Morphy, his calculated rating is about 2,300, which today would not put that person in the top one hundred players in the United States, much less the world. Paul Morphy was probably not as strong a player as Ken Regan is today. When I was a kid studying chess, I thought of Morphy as an idol (it's his best games that have been passed down), but arguably Morphy would match up about evenly with my own strength at age fifteen. That's still hard for me to wrap my mind around.

As players we are becoming more like the computers. Top grandmasters are more likely than before to experiment with "ugly" moves—or at least to give them further study—because now they understand that ugly moves are more likely to work out. Ken Regan suggests that human players are now more likely to understand the long-run value of sacrifices of material, and they are more willing to make such sacrifices even when the compensation for the sacrifice is nebulous rather than concrete. Humans are more likely to understand when it is possible to leave the King in an apparently exposed position, again because of experience playing with computers (Vasik Rajlich calls castling—removing the King to what is usually immediate safety—"the lazy move.") We've figured out that a lot of opening sequences are unsound—and how to beat them—and that a lot of opening sequences had been underrated. It's revolutionized our understanding of the game. What else will machine intelligence revolutionize?

There is a broader literature on the possibility of progress in general intelligence and not just among chess players. Average IQ scores have been rising for decades by about three points per decade, a phenomenon known as the "Flynn Effect." Of course it's not clear how much people, over time, have higher levels of general intelligence, and how much they are just getting better at taking the tests, but so what? Getting better at taking tests is still a form of cognitive improvement. (Most researchers in the area do in fact think that real intelligence performance is rising to some extent.)

There's further good news from the rapid progress of women playing chess. Bobby Fischer had famously joked that he could give "knight odds" (a significant handicap, which involves taking a knight off the board at the beginning of play) to any woman in the world. While this arrogant claim was never true, the relative paucity of top rank female players remained an embarrassment in the chess world until recently. Suddenly, something changed and the chess-playing strength of women improved remarkably. Judit Polgar has spent a good bit of time in the world's top ten and every male grandmaster fears her aggressive tactical prowess. Women are showing a long-term convergence toward the performance of men, and there is a boom of female interest in the game, including at the professional level. China and India in particular are producing some remarkably strong female chess players. Human beings can learn and develop talents long thought to be out of their reach.

What's interesting is that all the progress of female chess players doesn't seem to have a direct cause. To be sure, discrimination against women has been falling, broadly speaking, in the world for decades. In the chess world at some point there were enough female role models and enough female chess tournaments to support a growing female professional interest in the game. A few decades ago, chess was a realm where even the most committed gender egalitarians had to concede uncontested male superiority; "women just can't play chess" was the general opinion. The only change seems to have been that more women are doing it.

The Future of Intuition

Ken Regan once set out to investigate just how deep a chess game could be. He became interested in Grischuk vs. Kramnik, a well-known 2007 game from a major tournament in Mexico City. Working with chess engines, Ken tried to figure out if Grischuk had actually had a winning position near the end of the game. He ended up spending "the better part of a year" on the problem and ran more than ten trillion positions on his family's personal computer (his children still have standing instructions to be careful with open windows on the computer). The resulting file translates into roughly five hundred pages of analysis. The final judgment? In an incredibly rich and subtle position, Kramnik, with his best play, should have been able to draw. Ken, working with Fritz (the best program for the endgame in his view), generated a perfect understanding of the position, but that deep level of analysis is what it took and that was only one—admittedly interesting—position from a single game.

At the cognitive level, this unexpected depth is also a disturbing result. It shows that we humans—even at the highest levels of intellect and competition—like to oversimplify matters. We boil things down to our "intuitions" too much. We like pat answers and we take too much care to avoid intellectual chaos. Even if you don't think those flaws apply to everybody, they seem to apply to some of the most intelligent and analytic people in the human race, especially good chess players.

What does all this mean for our decisions, especially in the workplace?

1. Human strengths and weaknesses are surprisingly regular and predictable.
2. Be skeptical of the elegant and intuitive theory.
3. It's harder to get outside your own head than you think.
4. Revel in messiness.
5. We can learn.

The cognitive flaws illuminated by computer chess are not the flaws usually stressed by behavioral economics. First, as I mentioned earlier, behavioral economics doesn't always have a good standard for judging our rationality. Second, the behavioral economists themselves suffer from a lot of the same flaws that plagued the pre-computer grandmasters. They are looking for behavioral theories that are too elegant, too simple, or too intuitive, such as the abstract strictures of mathematical decision theory. For all their academic contributions, sometimes the nudgers are part of the problem rather than part of the solution.

If you test your ability to calculate against the truth of the computer, and continue to do so over the years, you'll get a lot better at calculation and you'll learn to transcend some of the natural tendency to rely on intuition. We humans are actually learning from the new technology, and that is a clear and deeply encouraging portent.

Vishy Anand, current chess world champion, noted, "Every decision we make, you can feel the computer's influence in the background." Garry Kasparov said, "Everyone looks at the position now from the computer lens."

It is both scary and exciting. Human intuition is becoming radically aware of its own limitations.

7

The New Office: Regular,
Stupid, and Frustrating

One of the biggest challenges in the new world of work is dealing with a world designed for the (relatively) smooth operation of machine intelligence.

Last year I had to call up my cable company, Cox, to get a line fixed, because it had fallen during a recent storm, and I feared it might cut loose altogether and sever service. You probably know the drill. In the old days you would call one number and a voice would answer. You would say what you wanted, and the voice would either answer your query or transfer you to another line, at which point you would speak to another voice. These days, you have to go through a menu of options, usually multiple times. I was asked what kind of service I wanted, with the questions branching into several parts, requiring several different buttons to be pressed on the telephone, each push preceded by a list of yet further options. You start reading on your computer screen while you are waiting for the machine to finish talking and you lose track of which button you are supposed to push, thereby requiring you to listen to at least one of the menus all over again, all of them if you are unlucky.

I had to enter my ten-digit phone number, even though I was calling from it and even though my phone service is through the cable company itself. I had to enter my zip code and I had to enter a four-digit PIN. Fortunately, I remembered mine (how many PINs or passwords do you have?). I then had to enter the first four digits of my street address.

Even then, I was not getting to where I wanted to go. I was being offered options that had nothing to do with "getting my cable line fixed." I lost patience, stopped pressing buttons, and simply repeated "operator" in a voice louder than was appropriate.

I finally could speak to somebody. But he was not some highfalutin' cable repair Einstein genius specialist; he was just a plain old ordinary guy. And I had to give him some of the basic information, such as the home address, all over again. He could at least tell me that someone would come by the next morning between 8:00 and 10:00 A.M. I survived the experience and the next day my cable was rehung.

Most of us have had experiences like this. But this is just the beginning. Your surrounding environment—the "help menu" of life so to speak—is about to get a lot simpler and a lot stupider and sometimes a lot more frustrating, all at the same time. I don't care if you're a fancy lawyer earning hundreds of thousands of dollars a year or more: You're going to be spending more time pressing buttons and completing parts of "the job" that the businesses just don't want to do anymore. And you thought spending money was supposed to be fun.

You may end up frustrated as I do, but these automated systems mean lower costs for the company and in the longer run lower prices for the customer too. Yet those machines are still going to need our help, and that means customers pitch in, often more than they would like. Self-service is another way of thinking about the dilemma of the modern workplace. Today's clerical or service worker has to compete not only against genius machines and off-shoring, but also against the customers.

You might wonder, by the way, exactly why these tasks aren't assigned to workers rather than customers. Why should you—the customer—have to work harder, especially in a world that is wealthier and technologically more sophisticated? The "fixed costs" of employing a worker—such as benefits and health care and training costs—are high and rising. If the task isn't that important, if the worker would often be idle, and if an assistant worker could not help the typical customer without some costly interaction maybe it makes more sense to put the burden on the customer, *provided that customer is teamed up with machine intelligence.*

Sometimes your task becomes harder or more frustrating when machines are put in a position to help you. It doesn't mean you are worse off for having the machines, but fairly often you will *feel* worse off, because the benefits of the new system are usually invisible yet the costs hit you in the face.

The Digital and Physical Environment

Let's start with a look at GPS systems. I am sorry to say that these devices still aren't all that good, even though they seem ubiquitous. This simple fact says a lot about where the service sector of our economy is headed.

Global sales of GPS systems were over forty-two million in 2011, plus it is expected that by 2014 there will be more than a billion GPS-enabled smart phones. That's a big and growing market, so you'd expect GPS systems to be pretty good; indeed, GPS systems use Einstein's general theory of relativity to keep track of the proper time, so they are not lacking in technical sophistication. Driving and finding directions is a much more significant human endeavor than is playing chess.

And yet, despite a lucrative market, GPS systems are a continual source of frustration. I've started using one, mostly for the purposes of research (and because my wife wants to). Okay, sometimes they're wonderful, but sometimes they're infuriating. On the plus side, I've learned a few counterintuitive but quite rapid routes that would not have occurred to me otherwise, and I've lived almost thirty years in my area (northern Virginia). I now think about driving more in terms of open possibilities and less in terms of a simple spatial map of how all the roads fit together; the latter approach can lead you to take the same routes over and over again, without wondering if there might be a better way. GPS, like Fritz and Rybka, has revised some of my preconceptions about what I thought I knew. The GPS systems also (usually) work well in apartment and condo complexes, when following seven steps of written directions, all specifying very quick turns in rapid succession, can be confusing.

That said, GPS doesn't do well at traffic circles or complex intersections, when "turn left" is utterly ambiguous and possibly dangerous. GPS doesn't understand the concept of a U-turn sufficiently well, or sometimes, if it makes a single mistake, it doesn't always get back on track. I've had it take me around a neighborhood or a block in a series of never-ending loops. It can't cope with a user typing in a wrong or incomplete address. There are more general reports, hard to verify, that GPS has caused hundreds of thousands of accidents (but how many has it prevented?). There are reports that it has stopped people on train tracks, sent them the wrong way on one-ways, led them to cross into flooded areas, and so on. Sometimes it distracts us from the road, or its voice of authority impels us to do foolish things, or we are not sufficiently meta-rational to know when to discard its instructions. Maybe it makes drivers ruder, as they push to get into the indicated lane even when they should be trying for the U-turn instead or maybe choosing a different destination altogether. One of my wife's cousins, who lives in Israel, claims that GPS can too easily take you into unacceptably dangerous territories—politically speaking, that is.

I'm happy to drive with GPS, but most of the time I want a map too, or Google or MapQuest directions, or some independent knowledge of the area.

Of course, Rybka hardly ever makes the mistakes of GPS. It's not that the GPS designers and programmers are incompetent; rather, Rybka has a more definite and better controlled environment. The "roads" of the chessboard are unchanging and unambiguous, and the pieces always move according to the established rules. It is never the case that a given square might be "closed for repair" or that a confusing traffic circle has an ill-specified turnoff. In short, when GPS fails, the problem is that human beings have not sufficiently reduced the complexity of the surrounding environment.

If you want a chance at beating a computer at chess, there is one simple way: Make the computer itself move the pieces across the board. For all the playing strength of the programs, the computer-driven robots that move the pieces don't do such a great job. At a 2011 industry conference in San Francisco a number of robots were tested on how well they could record the location of the pieces on the board and make the appropriate moves, as suggested by their accompanying chess-playing programs. It was not unusual for the vision systems of these robots to misread moves, pieces, and piece locations, and for their robotic arms to make the incorrect moves in response. Complicated environments—also known as the real world (even real chessboards)—are tricky.

Will we likely try to make our surrounding environments more like a chessboard and less like complicated, always-under-construction, multiple intersections? Imagine a world that is "stupider" than the world we live in today because it is designed more for machines, which require very literal readings of what is going on around them. It is a very real possibility.

The problem is that we humans are not always delighted by those very literal readings. We often prefer the vaguer and more personal world of calling up our cable company and explaining to someone, in ordinary language, what went wrong. We are used to hearing our spouses say something like, "I think you should turn pretty soon, maybe somewhere around here, what do you think?"—indicating many things at once and not all of them clear. When we are forced or induced to live in a world with more precision—but with more precise responsive interactions required of us too—we can get pretty grumpy. We can feel that our autonomy has been taken away or that we are being forced down rat holes, even when, all things considered, the new and more precise form of service is an advance. We feel like yelling "operator!"—to get out of this regularized, fully literal world. I see this problem only becoming more extreme.

To be sure, GPS will become better. It will be updated more rapidly and more effectively, and be able to read and communicate data about increasingly nuanced circumstances. This could be supplemented by the electronic dissemination of relevant information from the road authorities, much as electronic billboards serve some limited traffic reporting functions today, except the future data could be sent by wireless directly to the cars.

We see a bias toward regularized systems, rather than ideal systems, in the rise of the Kalashnikov AK-47, the world's most popular gun. It is not the technologically most advanced weapon, nor the most powerful, but it is easy to shoot, reload, and also fix. Sadly, you can give one to a child and have a working weapon rather quickly (as happens all too often during civil wars around the globe). Microsoft Word, in similar fashion, has succeeded because of ease of use and interchangeability, not because highly informed experts think it is the best software possible.

When it comes to GPS, strict regularization may not be possible anytime soon. When it comes to driving, it is not obvious that the gap between "the operating principles of humans" and "the operating principles of machines" will be bridged. And even if GPS driving becomes a totally seamless operation, intelligent machines will play a larger role in other spheres of life and we will still be dealing with these frustrating mixed systems. In the short and perhaps medium term there will be greater frustrations and some choices will be taken away from us. Your daily life will become a curious mix of both "much, much easier" and "more frustrating." It will be like living in a help menu. Overall, most people will prefer the new conveniences, but it is the frustrations that will stick in our memories. People will look back to a "time when life was easier," even though life will in fact be getting easier every year.

We've already made some of the adjustments. The ATM performs a lot of functions of the bank teller with obvious efficiencies. In turn we sacrifice some personal attention and the ability of a teller to answer questions about bank services. We're used to that one. It's a little more disconcerting when the machine swallows your card for no good reason. In our automotive lives, with the self-service gas pump, service stations carry hardly any human labor, so now it's harder to drive into a gas station and get your car fixed. It's a lot easier, however, to fill your car with gas. In part we've made that trade-off work because cars don't break down as much as they used to. Cell phones help too, because you can respond to an auto breakdown with a quick call home and to the AAA club or a tow truck.

More supermarkets in my area are using self-service checkout, and as a result grocery store lines have diminished, at least if you can do self-service. If you can't do self-service, the lines are longer because there are hardly any cashiers on the floor. It's also harder to ask someone where to find the dried cherries.

In the longer run the food will be cheaper because supermarkets don't have to pay so many cashier salaries. But try grabbing one kind of small green pepper, rather than another, and remembering the right pepper name at the checkout counter to track down the right code to punch in. I can't even figure out where to enter the code. You might stop buying the peppers altogether. When this economics professor goes to a self-service supermarket, he refuses to buy anything that has to be weighed, named, or otherwise evaluated on a discretionary basis. Only the fully standardized products, such as canned Goya beans and plastic containers of grapefruit juice, are purchased.

Like most of the problems I've been covering, this problem will someday be solved, and machines will solve it. I will be able to wave the green pepper at a camera and it will register the kind of pepper automatically.

The real conveniences will come as more and more interactions with computers come in the form of natural speech, à la Siri. Some-day I will walk into the house and simply announce to the sensors: "Cox cable company, I am home. Some branches fell at my house and now the cable wire is hanging dangerously low against some bushes. Please let me know times when someone could come by to fix it."

End of story, easy as that. But don't expect all the frustrations to go away all at once or to go away anytime soon. One theme of this book is that the advances of genius machines come in an uneven and staggered fashion. Just as Cox will get easier to deal with, intelligent machines, and the costs of coping with them, will become more prominent in other areas, such as our cars or our home appliances. For the foreseeable future, you'll always have to be learning something, reprogramming something, downloading new software, and pushing some buttons, all to have the sometimes dubious privilege of working with these new technological wonders.

Machine Assessments

It's not only our physical and digital environments that will become more regularized and more readily described in a literal manner. These same trends will happen to workplace personnel and our descriptions and evaluations of them. Workers will be increasingly tagged with their strengths and weaknesses, expressed in terms of numbers. The motive behind this development is essentially the same. Machines make literal and regular descriptions possible, and in turn such descriptions make machine analysis more acceptable and powerful. So we will try to make our workplaces more literal.

That will include having very exact readings on the quality of people we work with. The analogy with chess again proves useful. We all know how good a chess player is because each player has a numerical rating. Those ratings measure true quality quite accurately; "the sun got in my eyes" is an excuse that cannot be used very often. Chess ratings predict player performance remarkably well, with the exception of strong young prodigies who are improving their skills rapidly, or individuals with sudden health problems. These ratings are used for many purposes, including the decision of which players to invite to top tournaments, or how much of an appearance fee or lecture fee a player might deserve.

We can expect to have this practice spread more widely. The next step is to hire individuals to work with genius machines to assess the performance of workers, most of all skilled professionals. I mean the people we depend on, like doctors, lawyers, professors, and our coworkers too.

I asked Ken Regan if he could imagine using his research method—using Rybka to judge the quality of human chess players— to rate human performance more generally. In his view the average IT person would likely garner a 2,000 rating, comparable to that of a chess expert. He suggested that the cognitive performance of the average person would run to about a 1,600–1,800 rating, or equal to that of a good club player. As for newspaper writing, he graded the typical journalist at a strength of about 1,500. Ken was joking, but he had a serious point.

Machines aren't just about producing goods and services at lower cost; they also will improve the quality of service in the professions. Sooner or later, most professionals, especially at the top end of the market, will be graded by teams of skilled workers cooperating with smart machines. Think of this as more scientific Yelp ratings for almost everything, just as we now have such ratings for restaurants.

Let's say it is a lawyer. Potential customers can ask their smart phones where the lawyer went to school, what her class rank was, and what kinds of promotions she has received. That information will be accompanied by an asterisk: "This information explains only 27 percent of lawyer performance."

The better lawyers will open up their courtroom performances, their win–loss records, their contract analyses, and their written briefs to computer analyses for more accurate evaluations of professional quality. Siri will tell you: "This lawyer's written briefs are in the top eighty-first percentile of his peer group; that explains thirty-eight percent of performance on a corporate deal."

Many of the lesser lawyers will decline to be rated by a computer–human team at all, for fear of getting a bad rap and also because producing the rating will involve some cost. That will hurt their business prospects, especially with wealthier and better educated customers. Have you ever opened up the Friday movie page and read, "The studio declined to make this movie available for screening at press time"? The obvious conclusion is that the film is a dud, and indeed it usually is. They didn't do a screening because they wanted to avoid bad reviews. Some people do go see such movies, but they are usually the lesser-informed moviegoers, or those wishing to see a generic action or horror movie. In the broader context of professional performance, sooner or later most professionals will have to submit to ratings, one way or another, or bear the consequences and end up serving the lower and less informed ends of the market.

Will this development help or hurt the poor? On the down side, it will sometimes be harder to get a good doctor or lawyer on the cheap. Everyone will consult the published ratings and the best performers will charge higher prices. Still, on the positive side, these ratings will force professionals to compete more intensely against each other in terms of both price and quality. If you are a doctor in the bottom third, you will have to lower your prices and that will help poor people, who probably are not getting the best doctors anyway. We also can imagine ratings for "doctors with brusque bedside manners." These doctors might be cheap, even if they are good doctors, and the ratings will help us find them and save some money when we need to. Finally, the poor can save up and pay the high price when they really need a good lawyer or doctor, rather than paying high prices for subpar assistance, as is so often the case today.

The most likely outcome is that the poor will be big winners from these performance ratings. The well-educated wealthy already have fairly good means of finding good professionals, if only by asking their buddies, making donations to the local hospital, and pulling in favors from friends. This new system of ratings won't put everyone on an even par with regard to ability to pay, but it will equalize the information, and that should be good for most consumers.

The professionals may try to manipulate the ratings by taking on easier cases, such as turning away medical patients on death's door. Still, ratings production won't be based on all the cases a doctor or lawyer or teacher handles, but rather based on cases, patients, or other tasks that fit some standardized formula, so the intelligent machines can evaluate quality in a relatively fair manner. Other times the program will evaluate a professional by using a few summary, brute statistics, such as age, school attended, IQ, and income.

The numbers will be messy, but they'll come packaged with a caveat emptor: "This rating explains forty-three percent of the variation in performance quality." Such ratings are not an exact science, but the numbers themselves will reflect the imperfections of the systems.

Some professionals will rise to this challenge but many others will be demoralized. Just as the chess grandmasters no longer seem so wise and omniscient in light of computer analysis, so will doctors, lawyers, and teachers lose a good deal of their aura. They will be more like supplicants, waiting to be judged by the machines and usually appearing all too fallible, even at the higher ends of the market. We will end up with professionals who are less sanctimonious and less arrogant. Most doctors for instance are in fact B- or C-level players or worse compared to the very top, and this will become common knowledge. You will be less in awe of them as life-givers and think of even a "very good" doctor as somewhat akin to a third-string player who rides the bench at some lower-tier college.

It's going to be a very different world when consumers feel so much on top, and in some ways it will be more dangerous because consumers do not always know what they are doing. A lot of the problems in medicine are behavioral: Patients do not always follow the prescribed regimes of pills, exercise, diet changes, and the like. The word of the doctor is very often not enough, and I am talking about today's world, where doctors have this very strong mystique. Once professionals are rated, their customers and clients might scorn them more often and be less likely to heed their advice. A client may wish to plead "not guilty" when an experienced lawyer recommends the guilty plea instead, or recommends a settlement out of court. The client will bark back to the lawyer, "Look, you're not even in the top third of lawyers in Denver!" It will be harder for doctors and lawyers to "nudge" us and control us, because we will become more used to evaluating them, standing above them, and applying the programs to them in a manner that will make them feel small and will make many of us feel more powerful. If you pit me and Rybka in a match against Kasparov, I too would come away feeling just a bit heady, even if all I did was follow the Rybka recommendations move for move.

I've spoken of patients rating doctors, but doctors will rate patients too, especially if the doctor's own rating depends on how well a patient follows a prescribed regimen of treatment. How many doctors will want to treat the patients who do not follow up on instructions and take their medications? In the future we might see doctors turning away those patients, or charging them higher prices, or putting them last in the queue for a visit.

You've probably heard of FICO scores, which serve as credit ratings in the United States. The company that created FICO scores—Fair Isaac Corporation—is now working on creating a Medication Adherence Score, which is exactly what the name suggests. The company won't disclose the details behind the score, but it uses certain variables—such as how long a person has lived at one location and whether that person has owned a car—to measure the likelihood of taking proper medications. The current plan is to use this information to send vulnerable patients email reminders to take their medication, but of course such numbers usually evolve to serve multiple functions. Doctors will send away some of the very worst patients, in part to avoid wasting their time, in part to avoid the feeling of failure, and in part to protect their own performance ratings. There is, however, some bright side: We might end up with "patient quality adjusted" doctor ratings, which means some doctors will specialize in taking the worst and least responsible patients and doing whatever is possible.

What about medical privacy? Like it or not, in the future you probably won't be able to keep your quality rating a secret any better than you can today with your credit rating. Even if you did keep the "patient quality" information out of the clutches of your local doctor, the natural assumption will be that "failure to disclose" is a signal of irresponsible behavior as a patient. The good patients will choose to reveal their high performance ratings and the others will be lumped in with the bad patients if they do not reveal comparable information. Think back to those movies that "are not screened at press time."

How about a "consumer difficulty quotient"? Combine that with face- and gait-recognition technology. Useful information would flash on the smart phone of a salesperson in Nordstrom as the customer approaches the shoe section. The message might read: "Timothy O'Brien usually buys something rather quickly and doesn't waste your time," or perhaps, "Susan Boyle has returned one-third of her purchases in the last two years and complained twice about the sales help." Customers will have to live with the reputations they create for themselves.

One major current credit reporter, Equifax, is working on a "Discretionary Spending Index," which indicates whether customers (probabilistically speaking, that is) have extra money to spend. One goal is to sell this information to advertisers and direct mailers, but in an era of smart machines it will prove useful more generally, including on the dating market. It's already the case that 60 percent of US employers check credit scores before making a hire—this is a reality, not some dystopian science fiction world of the distant future. A lot of prospective partners will want to check the discretionary spending potential of the person they are about to date, or not.

Our (Legitimate) Fears

Ken Regan worries about our growing ability to measure and grade performance at a task, and what that means for an individual during the course of a career. Let's say he took his data and, retrospectively, established the features in a chess player's early games that predicted the ascent, or lack thereof, of a player. What if then a young player comes to him and asks Ken if he should continue at the game? Should that player quit a desk job and pursue chess full-time? Maybe Ken (with programs, of course) would be the one to know. But should he communicate his best estimate? On the affirmative side, it seems like he would be giving straightforward good advice. Don't we all use information to judge possible career paths and also to advise others?

On the negative side, too much knowledge can hinder achievement. What if a computer had graded Einstein at five years of age, when he still was not talking, and assessed his chance of becoming a great scientist? That truth, albeit some imperfect statistical estimate of the truth, will discourage too many people. Alternatively, maybe a future star will be done in because he is repeatedly told, from day one, that he is the anointed one. A certain amount of ambiguity may be good for the career ambitions of young people, and in the future we may miss some of the ambiguity we enjoy today.

Let us temper our emphasis on intelligence altogether. Kurt Vonnegut, in *Player Piano*, his prophetic 1952 novel about the future course of mechanization, put forward a different worry:

"Well—I think it's a grave mistake to put on public record everyone's I.Q. I think the first thing the revolutionaries would want to do is knock off everybody with an I.Q. over 110, say. If I were on your side of the river, I'd have the I.Q. books closed and the bridges mined. . . .

"Things are certainly set up for a class war based on conveniently established lines of demarkation [sic]. And I must say that the basic assumption of the present setup is a grade-A incitement to violence: the smarter you are, the better you are. . . . The criterion of brains is better than the one of money, but"—he held his thumb and forefinger about a sixteenth of an inch apart—"about *that* much better."

The character of Lasher speaks a bit more and makes it clear who, in his view, will do well in this new world of strictly measured value: smart people, but smart people with "special kinds of brain power. Not only must a person be bright, he must be bright in certain approved, useful directions: basically, management or engineering."

In other words, however useful the concept of standardization may be in the workplace, it can be scary when applied to social and economic relations as a whole. This is not a world where everyone is going to feel comfortable.

The discomfort with machines will expand along a number of dimensions. The most skilled man–machine teams will earn a lot, but there will be an issue of societal trust, precisely because their mastery of external environments may outstrip our ability to judge them. For instance, some top performers will stand above ready evaluation. We will know that they are very, very good, but it may be hard to monitor their quality with regular accuracy or ease, precisely because they are so capable relative to external standards of evaluation. We will rely on their goodwill and their morale to a high and possibly discomforting degree.

Their decisions will be difficult to second-guess. Consider the top man–machine medical diagnosticians, circa 2035. They will make life-and-death decisions for patients, hospitals, and other doctors. But what in a malpractice case should count as persuasive evidence of a medical mistake? The judgment of either "man alone" or "machine alone" won't do the trick, because neither is up to judging the team. Sometimes it will be possible to ascertain that a top human team member was in fact a fraud, but more typically the joint human–cyber diagnostic decisions themselves will be our highest standards for what is best. Having one team dispute the choice of another may indicate a mistake, but it will hardly show malfeasance.

When it comes to chess, no one seems to mind the existence of these very powerful man–machine Freestyle teams. They don't threaten powerful interests and thus they proceed unhindered. Will we treat all diagnostic teams the same way? Will we put absolute trust in them, with occasional checks for the fraudulent human interloper or with occasional detailed evaluations by other teams? Or will we regulate them or strike them down by applying cruder standards of evaluation to their decisions, such as subjecting them to "man-only" judgments?

The way our legal institutions are currently set up—and I predict they won't change soon—we will be judging Freestyle diagnoses by applying the standards of man alone. Our legal systems are based on jury and judge decisions and they don't involve a lot of computer use, except through the medium of expert witness testimony. It is not general practice for a judge or jury to switch on some computer programs for advice. Imagine a judge punching in five key features of a case, being told by a genius machine that the chance of guilt is 67 percent, and then pondering how to interpret the confidence intervals that are served up. We are pretty far away from accepting that world as citizens and voters, even though it may be technologically feasible in the near future or even now. In the meantime, we will have judges and juries applying their own wisdom, or lack thereof.

Often the diagnosticians will appear to be off in their judgments, but we know in broader terms, when we judge the matter from some distance, that the teams are in fact the strongest evaluators. In response to pressures from evaluators, sometimes the teams will ensure that their actions are defensible to their human evaluators rather than make the very best decisions. When it comes to medicine, a brilliant risk might be passed by for fear of having to take on the blame.

By their nature, Freestyle decisions by the best teams do not allow for easily replicable means of judgment and evaluation; if they did, the machine itself would have already embedded such judgments in its algorithms. Freestyle decisions at the highest levels are again—by their nature—uniquely creative acts. And it seems those uniquely creative acts often scare us or intimidate us or make us feel uncomfortable. It means that someone out there is able to act without facing much accountability.

In the language of today's financial markets, a lot of financial trading occurs "off the balance sheet," meaning it faces different reporting requirements and it stands outside many financial regulations. Sometimes regulators or commentators are upset or outraged over this fact and indeed it is easy to see the problem. Yet it is not so easy to insist that the trades and positions be put "on the balance sheet" in any easily interpretable manner. Trillions of dollars of open positions in derivatives, mixed in with complex computer-driven trading strategies, just doesn't fit in too well with the ideas of transparency and accountability.

In times of war or military action, we will end up relying on machines, as embedded in drones or long-distance surveillance, to decide when and whether to kill a target. Is that an Afghani wedding party or a group of terrorists? What if the chance of terrorists is 37 percent and there are seven innocent children accompanying the group? It is easy enough to imagine equipping a machine with the information to calculate the best probabilities available and to make such decisions "on the spot." I am not sure they would be morally better decisions than the humans would make but at least they would be better from the point of view of the commanders. The only question is when this will happen. Yet there is again the tricky question of how to judge mistakes and what a disciplinary action or war crimes trial for such actions might look like. Are the programmers of the machine liable? The generals who deploy such machines? The president of the United States? How is a machine mistake different from a gun that misfires and kills some innocent civilians? My best guess is that the powerful nations, such as the United States, will simply let these machines operate without any constraints beyond what is embedded in their original programming. Mistakes, including moral mistakes, will be ignored and covered up, or possibly never rooted out in the first place.

We are again back to that scary world without much accountability for some very important decisions.

In other contexts we will not be able to avoid making some very direct moral judgments. Let's say for instance that you own a driverless car. How should such a car be programmed in case an accident is imminent? Should the car swerve away from hitting a baby carriage, but at the risk of running into two elderly people? Should the car be programmed to crash you into a telephone pole, rather than run a $p = 0.6$ chance of knocking over a pedestrian? It is easy enough to imagine such issues being debated on the evening news. I don't imagine that public debate will let the programmed cars behave very "selfishly." But what if a human driver takes over operation of the car right before the crash occurs? Currently the legal system allows a fair amount of leeway to human judgment. Might our legal standards for human drivers toughen up too? Will the moral calculus of your "driving program" be admissible evidence for your reckless-driving hearing? Probably so. You can see that a lot of everyday morality is going to change.

Overall there will be some systematic problems when machine intelligence is judged by external and highly imperfect standards. What is it like when the superior diagnostics team is forced to satisfy the standards of a cognitively inferior boss, governor, or legal system?

We're going to find out. The future will bring to us The Unaccountable Freestyle Team, The Scary Freestyle Team, and The Crippled Freestyle Team, all at once.

8

Why the Turing Game
Doesn't Matter

It's the bumps and delays that will make the rise of smart machines a livable process. It could well be destabilizing if the technologies of mechanical intelligence—two hundred years' worth of progress, plus their complementary applications—were placed in our hands overnight. A lot of people couldn't get any jobs whatsoever because they couldn't work with the advanced machines, and it would take them a long time to learn. We deal with machines today as well as we do because our progress has been gradual, allowing us to learn along the way. When it comes to technology, progress is usually good, but gradual progress is usually better.

If I were suddenly transported back to medieval times, or even to the nineteenth century, I would be pretty hopeless and probably unemployable. I wouldn't know how to shoe a horse or work a water wheel or light a lamp with whale oil or conduct a church service in Latin. Eventually I would learn some of these skills, but it would not be easy going. Injuries from the hand tools would be unnerving and the absence of electrical power frustrating. I'll have a much easier time handling the iPhone 6 or the next generation of computer software, even though I would describe myself as "bumbling" in my digital tech skills.

The Long Run

Not everyone sees the path forward as gradual, and there are some radical visions of how mechanical or artificial intelligence will change the world. Eliezer Yudkowsky, the well-known futurist and speculative thinker, offers a bleak prediction: We wake up one day and find that a super-intelligent machine has taken over the world, as happened in the Arnold Schwarzenegger movie *The Terminator*. Perhaps this machine will destroy or enslave us. Interestingly, in Ambrose Bierce's early and seminal fictional account of a chess-playing computer, "Moxon's Master," from 1909, the machine apparently turns murderous.

Eliezer fears a cascade. Once a very good program becomes sufficiently developed, it will create other programs, which in turn will create other capabilities. The combined ability of all these programs could explode exponentially and, supposedly, have a lot of power. Recall the motto associated with the Terminator movie series: "Skynet goes live." It was downhill after that, even though they managed to extract a few sequels about subsequent battles.

The result would be that we would all work for one machine or another, not necessarily voluntarily.

If we're talking about the distant enough future, we must assume virtually anything is possible. But the evidence so far doesn't suggest this kind of unstable cascade to be very likely. Even the strongest programs need human assistance every step of the way. Chess programs for instance outperform humans at narrowly circumscribed tasks, such as calculating chess variations. At the very least the programs still need humans to turn on the computer and start the game, not to mention organize the tournament or build the opening book. No one fears that the programs will stage a coup and take over the United States Chess Federation. The idea sounds, well, silly. The programs show no signs of "thinking for themselves," and their most brilliant "intuitive" moments are the result of deeper pure calculation, not an ability to formulate independent creative plans or to introspectively reflect upon their love or hatred for their human chess opponents. For all their practical abilities, there is no reason to expect these programs to move down the corridor of self-awareness, and it is easy to see that they operate under quite different principles than do human brains. The truth is there are no real vampires, no dragons, and no HALS. Let's not worry about them appearing under the bed or in our hard drives.

The history of the chess programs also shows that progress is gradual and piecemeal rather than a result of exponential capabilities that explode overnight. Computers are still struggling to master the more complex board games of Shogun and Go. There too, computers will improve, but with human assistance and leadership. No one has noticed Rybka spontaneously teaching itself these games in its spare time.

In another dystopian vision, the proliferation of computer intelligence will bring about a Malthusian world where human laborers will struggle to earn subsistence. Economist Robin Hanson, my colleague at George Mason University, considers this scenario in the fascinating and influential paper "Economic Growth Given Machine Intelligence."

Imagine a limiting case where smart machines can do all the tasks of humans. The economics of this situation will prove problematic. The wage rate then cannot exceed the cost of producing a competing machine, because if it did no one would employ the human. Eventually the machines will become fairly cheap to produce, so human wages will fall correspondingly. It's even possible that the machines will be cheaper than the level of subsistence wages. In that case either workers must live off charity or the population will shrink rapidly or some mix of both. One thing that Robin shows with his model is that you can have a long period where—for reasons of complementarity—machines boost wages, but eventually machines substitute for intelligent labor and wages can fall rapidly.

Malthusian wages do not mean an impoverished existence for everybody, of course. The machines are still owned by someone, and the owners of machines are very wealthy since the machines can produce a lot of goods and services very cheaply. If just about everyone has a stake in the machines, this could be a utopia rather than a dystopia. Alternatively, perhaps the government owns a share in the machines and it uses that wealth to support the remaining poor, who did not buy machines in time and who now cannot find jobs because of competition from the machines. They will become wards of the state, much as many people live off of oil wealth in some of the less populated petro-states.

These extreme cases can help us identify trends, such as downward pressure on many kinds of wages (as discussed in earlier chapters with regard to the market for manual labor). Robin's paper is thought-provoking. Still, I am looking across a more modest time-scale and a more modest set of changes. Robin's analysis *might* apply to the very long run, perhaps hundreds of years from now.

But for the next fifty years or longer, the Freestyle model is more applicable. Most AI applications still require human support, and those applications, even if they spread considerably, will not come close to displacing all human jobs. Instead, intelligent machines will replace some laborers and augment the value of others in a slow and halting manner.

The most radical hypothesis about future technology is Ray Kurzweil's vision of a machine intelligence "Singularity." Kurzweil argues that mankind will obtain the capacity to scan brains and upload them into computers. There will be many copies of each "person" and presumably these entities will exist for a long time, with the multiple copies making the "person" hard to wipe out, even in the event of a system crash. I've heard some of Kurzweil's followers claim this scenario will happen within the next fifty years, and Kurzweil's writings seem to encourage such speculations. Arguably the uploaded entities could be enhanced, so they might combine the best features of human and machine intelligence.

I suspect this will never be viable, if only because the human brain is so intimately connected to the human body, and because it so relies on the body for inputs and nourishment. For instance, scientists are learning how much our brain relies on our stomach ("thinking with your gut" is closer to the truth than we used to believe) and how much our brain relies on the more general interactions with our bodies and the external environment for its processing capabilities. Moving, and interacting with the environment, is needed to set in motion, sustain, and enrich our thoughts. That means "brain emulation" requires building a whole working body (or significant parts thereof), not just an abstract, digitalized "brain in a vat."

At that point, anyone might wonder whether it isn't easier to start with the bodies and brains we already have and make them more effective by allying them with machines, or using machines as add-ons. The Freestyle model seems a lot more economical, and to most people a lot more palatable, than Kurzweil's utopian project of brain uploads. Economic incentives, namely the drive to produce effective man–machine teams, will direct the attention of innovators to produce machines that complement human intelligence, not machines that capture or mimic it. The demand for immortality may lead some to seek out digital copies of themselves, but to most people this maneuver doesn't counter your death any more than Xeroxing your passport or maintaining your homepage after you have passed away. Many people do in fact keep their webpages going, but that kind of indirect immortality does not have a strong enough psychological pull on enough people to attract a big enough chunk of research and development dollars. I won't miss it myself.

Should we really think that scientists will inject thousands of mini-bots into our brains, each equipped to take very accurate photos of specified areas, and turn and integrate those photos into some kind of information-preserving copy, all before the Chicago Cubs are due to win a National League pennant again? I say no.

All of these extreme scenarios, whatever their differences, share some features. They are about worlds that we cannot today control, influence, or even visualize clearly. They tend to be about worlds we cannot investigate empirically or find direct parallels for today. They are somewhat like religious allegories, involving some mix of concepts of deliverance, resurrection, destruction of the current world, and immortality. Given the appeal of such tales, I'm suspicious that these extreme scenarios are living off of their emotional appeal, and they are a kind of religion for computer nerds. Let us set that devotion aside.

Man–Machine Convergence

Okay, say we're not going to become uploads. But is there a convergence, even a behavioral or a mimetic one, occurring between man and machine?

Some commentators attach great weight to whether intelligent machines can pass a Turing test. If a group of intelligent observers were conversing with a machine through an exchange of typed messages, could those observers tell whether they were talking to a human being or to a machine? If the machine can consistently "pass" as human, it has succeeded at the Turing test. Sometimes the Turing test is considered to be a benchmark for when we can call a machine truly intelligent.

So far the machines have had only mixed results at passing the Turing test. Getting human conversation "right" is a complex, multidimensional problem, highly sensitive to context and background understanding, and so it isn't easy for a computer program. Here is a dialogue that Teppo Felin, a sociology blogger, had with Rosette, the program that won the 2011 Loebner Prize, which is dedicated to the Turing task:

ROSETTE: Hello, again. Did you have a good day?
TEPPO: congrats on winning the Loebner prize!
ROSETTE: Why are you so excited about this? I think shopping summons up the ancient hunter-gatherer in me. I like having a mission to find something I need or want. Sometimes I'll research online, to find a great deal, but I like rummaging through Jeremy's up the road, if I want an outfit for a special occasion, or just for fun.

TEPPO: did you win the prize or not?

ROSETTE: Not often. do you?

TEPPO: Hmm, maybe I am talking to the wrong Suzette, though the web page indeed does confirm that you are the winner.

ROSETTE: Wow. My housemate is also named Suzette. She's from Georgia. Where do you live?

TEPPO: Hold on. So, again, did you win the prize or not?

ROSETTE: Could you rephrase that in a more positive way? What kind of music do you like?

TEPPO: Ok, I give up.

ROSETTE: Who's your favorite group or musician?

TEPPO: Sorry Rosette. I can do a better job of pretending I think you are human than you can pretend to be human.

There are more successful attempts. At a recent tech festival in India, Techniche 2011, a program called Cleverbot competed on its conversational skills. The audience watched a bunch of conversations, some computer–human, and some human–human, without knowing which were which. Of the voting audience, 59.3 percent judged the computer program to be a human, whereas 63.3 percent judged the conversing humans to be human. Maybe those voters weren't concentrating hard enough on the differences, or maybe the interrogators weren't trying hard enough to trip up the machines. It's also the case that the computer probably does best in a radically polyglot society, such as India, where dozens of different languages are spoken, malapropisms are relatively common, and fluency in a particular language is not taken for granted. Still, those two vote totals are pretty close, and arguably these programs have almost passed the Turing test.

However, once you take a "gains from trade" perspective on man–machine interaction, the Turing test seems to be more of a curiosity than a fundamental standard for judging the presence of intelligence or not. What good does passing a Turing test really do?

There has been an enduring misunderstanding that needs to be cleared up. Turing's core message was never "If a machine can imitate a man, the machine must be intelligent." Rather, it was "Inability to imitate does not rule out intelligence." In his classic essay on the Turing test, Turing encouraged his readers to take a broader perspective on intelligence and conceive of it more universally and indeed more ethically. He was concerned with the possibility of unusual forms of intelligence, our inability to recognize those intelligences, and the limitations of the concept of indistinguishability as a standard for defining what is intelligence and what is not.

In section two of the paper, Turing asks directly whether imitation should be the standard of intelligence. He considers whether a man can imitate a machine rather than vice versa. Of course the answer is no, especially in matters of arithmetic, yet obviously a man thinks and can think computationally (in terms of chess problems, for example). We are warned that imitation cannot be the fundamental standard or marker of intelligence.

Reflecting on Turing's life can change one's perspective on what the Turing test really means. Turing was gay. He was persecuted for this difference in a manner that included chemical castration and led to his suicide. In the mainstream British society of that time, he proved unable to consistently "pass" for straight. Interestingly, the second paragraph of Turing's famous paper starts with the question of whether a male or female can pass for a member of the other gender in a typed conversation. The notion of "passing" was of direct personal concern to Turing and in more personal settings

Turing probably did not view "passing" as synonymous with actually being a particular way.

It has also been speculated that Turing was autistic or had Asperger's syndrome, which suggests his mind was of a very different nature compared to most of the people he knew. Turing probably was not aware of these neurodevelopmental concepts as such (they had not yet entered common usage), but surely he knew, growing up, that he was in some ways very different from others. In public school he was judged to be "ludicrously behind" with "the worst" writing ever encountered, and he was singled out as "bound to be a problem for any school or community." Whether or not he was autistic, it is clear from published accounts that other people noticed he thought and acted in highly atypical ways. Turing failed to imitate the people he met in mainstream British society. Of this he must have been acutely, painfully aware.

Turing does not make this point, but many human beings, intelligent and of mature age, could not pass what we now call a Turing test. This includes many human beings who would do well on IQ tests or other traditional measures of intelligence. Some autistics provide examples, as would other individuals with nonstandard neurodevelopmental paths. For instance, an autistic might not comply with the presumed or implied typical social context behind many of the proffered questions. Many answers from autistic individuals might seem "off" to the judging panel, so there is a good chance autistics, who have distinctly atypical methods of information processing, would fail to pass the imitation or Turing test.

This point about the limitations of imitation as a cognitive standard isn't relevant to autistics only. In the 2011 Techniche tests in India, only 63.3 percent of the audience could tell that the humans were indeed humans, so what were those people "doing wrong"? Once the question of humanity comes into doubt, we start second-guessing the human interlocutors and realizing how idiosyncratic they are and then we wonder whether they are really humans after all. Is that funny response an endearing quirk or a computer glitch? How the dialogue seems to you will depend on the presuppositions you bring to the evaluation.

It's interesting to see *how* Cleverbot convinces a lot of people that it is human. It doesn't try to respond logically to every query, or to correctly parse every sentence from the human. Instead it tries to mimic how an incoherent or slightly crazed human would converse—or perhaps how a pickup artist would try to dazzle a sweet young thing with banter—as its best chance of doing a passable imitation. Here is a snippet from one dialogue:

OK, I imagined them, now what?
CLEVERBOT: I didn't mention anything I don't like.
If people in society don't contribute to making society better,
    why should they be allowed to live in that society?
CLEVERBOT: But there are many other worlds.
Then why not make them go to the other worlds?
CLEVERBOT: I will stain my hands with your blood.

At best you will think that is a human who just spent some time smoking pot, hardly what the program set out to achieve. Go to the website itself, Cleverbot.com, and see what you think. There you will be warned that the program may serve up unusual, unexpected, or inappropriate content. The first question I posed it was, "Can you pass a Turing test?" The answer "What is a Turing test?" was quite human, I thought. Then:

TYLER: A Turing test shows whether you can mimic a human
        interlocutor.
CLEVERBOT: Do the Turing test on me.
TYLER: So far you are passing it.
CLEVERBOT: Ah well, what can one do?

Not bad. Yet then I wrote, "Jump off the Brooklyn Bridge!" and
it lost the test when it answered, "But that magic jump on me."

The historical evidence, while sparse, suggests that Turing himself commonly served up unusual responses in dialogue. One BBC
radio producer doubted Turing's ability to communicate and complained about the "definite hesitation" in his speech. There is also a
transcript of a discussion between Turing and several other scientists; one commentator observed: "Reading the transcript is rather
like reading the conversations generated by computers. . . . Few of
the discussions can stick to a point or actually address a question!"

It would be funny, and appropriate, if the first computers to pass
Turing tests did so by, at least in part, mimicking the conversational
style of Turing himself.

What does this say about the collaboration between humans and
computers? Convergence isn't needed. In his quest to take down the
notion of imitation as the standard of intelligence, Turing notes
that a machine will be most effective, cognitively speaking, when it
does something *other* than imitate or try to imitate a man.

Someday a machine will pass a Turing test, and it's probably not
far away. That day will be a nonevent at first, although slowly its
ramifications will become apparent: the competition for your attention will become more fierce.

Take online dating. It will be possible to undertake conversations with all desired dating profiles, through the Turing-worthy

programs, of course. One's "bots" will be able to send inquiries to many eligible potential partners, and in turn the bots can respond to all profile inquiries. After some exchanges, the genius machines will judge how much the respondent "gives good email." Or perhaps it will be my bot judging your bot in a bot-on-bot email dialogue. But if it's you who created your bot, maybe that's good enough. If my bot likes your bot I might like you too, a bit like how park encounters are sometimes mediated by how well the dogs of the two parties get along. We'll see soon enough, when our bots tell us if that correlation—between "bot liking" and "human liking"—is going to hold up.

There are, by the way, studies of which kind of online approaches are most likely to elicit a positive response. One linguistic analysis indicates that it is best when the approach does not use the word "I" too many times (you may recall from chapter one that this same variable indicates the writer is more likely to be lying), uses the pronoun "you" frequently, avoids usage of leisure words such as "movie," and uses more social words such as "relationship" and "helpful." Perhaps contrary to the intuitions of many, the use of negative words did not render a response less likely.

Or say you're a professor holding online office hours: At what point will the students prefer to pose their questions to the bot? Gmail chat, or whatever takes it place, will be more of an adventure because you won't know who is really responding. Online therapy—or therapy in Second Life—is already popular, and at some point the therapist may disappear. Perhaps a lot of the effect of therapy was a kind of placebo anyway, but will it be equally satisfying to know that a computer is commiserating with you and telling you that many other people share your problem? It's easy enough to imagine unethical therapists who log on first as a person and then turn over the talking to their bot.

There will be no end to the number of media and marketing pitches, because the initial query will be backed by a responsive, Turing-worthy program. Or set up your bot to go around asking for free products, free favors, free whatever, all sent by email, knowing that the bot can handle the back-and-forth response just fine.

In short, there will be no end to the number of questions that are asked and answered once the Turing test is cracked, especially once it is cracked for a wide variety of situations, not just for anonymous one-on-one chats. This proliferation of questions may force real people out of the business of asking and answering questions, because they will be swamped by a kind of conversational spam. That in turn will create even more space and opportunities for the bots. "Face time" will become all the more important as a signal of actual interest and caring, because "computer time" will be too easy to replicate through the bots. Maybe you'll use Skype to prove it is really you, and that will work for as long as bots cannot replicate your facial expressions and voice patterns through a streamed image.

These days, we're even finding computer programs that can pass aesthetic Turing tests, so to speak. Computers are composing music, and it's not always easy to tell which tune comes out of a human and which comes out of a computer. Computers not only play chess but now judge the aesthetic qualities of various chess problems and compositions. Or consider the robots that produce sketches of human faces. On the next page are two examples, one by a human artist and the other by a robot (the left is created by a machine and the right by a human).

When I first saw those two images, I had no idea which one was the product of machine intelligence.

Sometimes the skills of computers will be used for outright cheating, and we can expect this problem to increase as the computers get better. In 2011 a number of French players—strong but not world champion class—were accused of cheating during their tournament chess games at the Chess Olympiad. These

© Patrick Tresset

individuals were Sébastien Feller, Cyril Marzolo, and Arnaud Hauchard, and they allegedly used mobile text messages, a remote chess computer, and coded signals to relay the critical information. Subsequent computer analysis showed that they were playing some of the highest-quality chess of all time, comparable to that of a top program. That does not militate in favor of their innocence, and the players have been suspended from major tournaments. It did not help when an outsider stumbled upon a text message sent by one of the ring members to Mr. Marzolo, which read, "Hurry up and send me some moves."

Some chess players may carry Pocket Fritz (it can be used on an iPhone and is still strong enough to help a grandmaster) and sneak off to the men's room to consult with the little daemon more than is appropriate. Other players may use an accomplice to watch a tournament game and relay moves back to a computer program. The accomplice, after reading the computer recommendation, revisits the playing area and sends signals about the next preferred move that only the player can pick up. The signals themselves could be anything: how and where the observer stands, how many times he scratches his head, and so on. It's easy to imagine how such a code could be generated, although we don't know how often such strategies are actually executed.

The cheating once took a very different form. In the eighteenth and nineteenth centuries, the trick was to sneak a human chess player inside a machine and pretend to have created a technological marvel—a machine that played a good game of chess. This formed the basis of a sensational traveling exhibition called a Mechanical Turk, which hid a human inside—in a nontransparent manner—using principles now associated with magicians. (If you are wondering, the name of Amazon's current Mechanical Turk service, which combines man and machine to perform programming tasks, is based on this history.) The machine "operated" from 1770 until its destruction by fire in 1854, although it was exposed as a fake at least as early as 1820. It was originally designed to impress Queen Maria Theresa of Austria, and the contraption is said to have defeated both Benjamin Franklin and Napoleon Bonaparte at chess. The machine also toured the United States and Cuba. The modern switch is that now the cheaters are humans who sneak the machine along, not vice versa.

The Mechanical Turk contraption was used to cheat on an early version of the Turing test as well, although it was the human inside who was doing the thinking and answering. Spectators were asked to come up and present the machine with questions and see if they could converse with it as they would with a human. The machine had an artificial hand, and the human inside could use that hand to indicate letters on a board, thereby spelling out words and, with some patience, entire sentences. Crowds marveled that a machine could respond as a human would. In a strange yet fraudulent anticipation of Turing, Wolfgang von Kempelen, the original designer of the machine, realized that this conversation ability would be the machine's most impressive function of all, even more impressive than winning at chess. (Von Kempelen also built a machine that, through use of hoses and tubes, could imperfectly replicate some sounds of the human voice and thus some words, and that was without any human inside.) Spectators were witnessing an early-day version of Siri on the iPhone, albeit one based on fraud.

It's still quite difficult, however, for the human to use the machine at chess and still appear to be playing like a human. Ken Regan's work, which I discussed in the chapter on behavioral economics, provides a way to catch computer cheaters, or at least pronounce them culpable, statistically speaking.

As it stands, if a player makes too many moves that coincide with the computer's recommendations, suspicion is thrown on the player. Indeed, there have been cases of false or even possibly libelous charges having been thrown about. In the 2006 world championship match, Veselin Topalov vs. Vladimir Kramnik, Topalov's manager, Silvio Danailov, claimed that Kramnik was visiting the men's room too often and somehow receiving computer aid; it's now called Toiletgate. This charge turned out to be groundless and more of a tactical ploy for psychological disruption than anything else, to the enduring disgrace of Danailov. Eventually Kramnik won the match above all suspicion, beating Topalov in a sudden-death, stay-at-the-board, blitz match. Still, the issue will not go away, for obvious reasons.

Ken Regan stresses that play that matches computer recommendations should not suffice to convict a player for cheating. In some chess games, a lot of the moves are forced or force the hand of the opponent; in those cases, relatively good players will agree with the computer a lot of the time because both human and computer can see that there is only one good move for an entire series of positions in the game. It's also the case that strong humans "play the computer move" disproportionately often in relatively straightforward or simple positions. Strong players with a straightforward style sometimes appear to be cheating when the only thing going on is strong play in simple positions. In fact, if you simply filter the data, without interpreting it intelligently, one of the biggest "computer cheaters" of all time is possibly Jose Raoul Capablanca, the Cuban grandmaster renowned for his clear and simple style—in the 1920s and 1930s! In other cases, some of them coming before the invention of the quality programs, there is an outlying, extraordinary human performance at the far end of the distribution of quality, as indeed you would expect when so many games are played over the years.

That all said, a true systematic cheater will usually leave telltale marks on his or her games, at least if the cheating is repeated rather than one-off. Ken's method is likely to catch the chronic crook, but it will not detect a grandmaster who cheats only at one critical turning point in the contest.

In any case, rather than converging, man and machine are likely to become more different in some ways, including cognitively. Most of this book is about the evolution of the machines, but people will change too. I'm not talking about longer-run changes in the genetic code, but rather more simple changes in how we live our lives and which skills we decide to acquire or not.

To put it bluntly, we are outsourcing some parts of our brain to mechanical devices and indeed we have been doing that for millennia, whether it be to writing implements, books, the abacus, or a modern supercomputer. In response to all of these developments we have focused more on the skills that the machines can't bring us.

Memory and "Search"

When I was a kid and needed to know some facts, I would call up the reference department of the public library in Hackensack, New Jersey. I still recall their number, which I had memorized: 201-343-4169. But those calls were a pain, and often the library staff would have to call me back after finding the right answer by going through the card catalog and Dewey decimal system and scouring the library shelves. We don't do it that way anymore, instead preferring to start—and often finish—with Google or other methods of search, such as Twitter or apps. These technologies are marvels, for all the reasons that by now have become clichés. We also don't memorize so many phone numbers anymore, instead preferring to enter them into our cell phones or relying on Google to call them up when needed.

The Google crutch, if I may call it that, influences how we think and how we learn. There's now good systematic evidence about how Google changes our mental capacities, and I think most of us have experienced this personally as well. When people use Google more, they lose some of their ability—or at least willingness—to remember facts. After all, why should you keep track of all that stuff? If it is a factual question, the answer probably is right at your fingertips, especially with smart phones and iPads. In similar fashion, it seems that people who manage accounts became less skilled at some memory functions once they obtained cheap paper, writing instruments, accounting books, and other means of keeping track of figures.

In earlier times there was a prominent "science of memory," in which was taught the skills of remembering numbers, people's names, sequences of numbers, and so on. This science goes at least as far back as ancient Greece, and it flourished during medieval and Renaissance times. Journalist Joshua Foer recently wrote eloquently about this kind of memory discipline in *Moonwalking with Einstein* and returned some prominence to it. He focused mostly on one particular technique known as the "method of loci," by which elements you wish to remember are given a position in an imaginary place along with some surprising or colorful attributes. Foer recounts his adventures in the USA Memory Championship using his newfound skill of memorizing decks of playing cards.

The ancient arts of memory, in their most general form, are techniques to improve your mind. These arts were not just about memorization and many of their advocates drew an explicit distinction between the memory arts and memorization. The memory arts were about learning how to order ideas in new ways, and thus the memory arts were a path to composition and innovation and the generation of novelty. It was about taking older and simpler parts and from those parts making new things, be they hymns, poems, prayers, books, or a new appreciation of the wonders of God.

The point was to make your ideas "searchable," as a computer literate person would say. That is, you could start with a question or some simple starting points and be led, by the algorithmic nature of the memory arts, to more complex ideas and truths. When it comes to Google, the right magic keys can get you to many new places, and those keys consist in the relatively manageable art of knowing the right search words. Google is the successful embodiment, through technology, of the earlier dream of the memory palace.

For many centuries the idea of an algorithmic path toward greater knowledge was an obsession in Western thought and religion; it infused the Kabbalah, many of the medieval scholastics, and scientists such as Isaac Newton and Johannes Kepler. It fell out of favor as it was increasingly regarded as ridiculous, but guess what? These visions made perfect sense but just didn't yet have the right technologies to make them work.

For a long time the memory tradition in Western thought appeared to be a dead end and indeed few people today use memory theatres or other memory tricks; the technology never seemed practical for most of us. Frances Yates's 1966 book *The Art of Memory* traced Western thought on memory up through the seventeenth century. Yates's work is famous among scholars of the Renaissance, and it has also resonated with people who study Kabbalah or Hermetic or mystical themes in Western history. Her work as a historian of ideas received rave reviews, but most scholars thought of it as a curiosity, albeit an interesting one. Now we're seeing these early writers for what they were: brilliant precursors straining after some of the most important ideas and techniques of the contemporary world, in this case the art and science of search.

Returning to the present, Google is making a lot of the memory arts fall away altogether. Nonetheless, that doesn't mean, as critics such as Nicholas Carr have alleged, that we are becoming stupider.

First, we presumably learn something useful through Google, and that information also gives us broader background knowledge for understanding and interpreting other facts about the world, whether they come from Google or not. Second, we have become much better at searching for answers, and that too is a skill. Rather than remembering a fact, I often remember how I can best search for a fact. A lot of my searching is done through my blog, which catalogs parts of what I know, and my Gmail account, where I store useful information. Where am I having lunch with Steve Teles tomorrow? I don't remember, but I do remember that I ought to search for "Steve Teles lunch" in my Gmail account and I will arrive at the right answer very quickly. I also have developed a good sense of when it is better to search through Google and when it is better to search through Twitter; for instance, search through Twitter when you are looking for rumors or for very current information, say within the last half day or so.

From what I see, most people prefer to give up some memory to enjoy this symbiotic relationship with modern search.

Which other parts of our brains will we outsource to intelligent machines? And how will we change as a result? It's already clear that the genius machines do very well at brute calculation, so the obvious prediction is that we will, as humans, become less interested in pure calculation and less able to calculate. That's already happened. The pocket calculator has weakened our ability to perform seamless long division with pencil and paper, but we spend our time honing other skills instead.

Two different effects are operating here, but we can tease them apart for a look at where humanity is headed. On one hand, many successful individuals will learn how to think like smart machines, or at least enough to understand their operation, in order to become wealthy, high-status earners. In that way we will become more like computers—well, a large number of high earners will become more like computers anyway, cognitively speaking. That said, when it comes to our private lives, we will become less like computers, because we rely on computers for many basic functions, such as recording numbers, helping us with arithmetic, and remembering facts through internet search. In these ways we will become more intuitive, more attuned to the psychology and emotions of everyday life, and more spontaneously creative.

The Machine's Place

Not many people, from their standpoint as observers or as consumers of culture, want man and machine to converge. It seems that even when the machines are a lot better than the human competitors, we're not so interested in watching them.

For instance, when it comes to chess we humans don't seem to care when the machines play each other. Hardly anyone is watching or talking about the computer vs. computer games. The online commentary is a trickle, most chess websites—even those for chess specialists—don't bother to reproduce the games, and I never see them reported in the popular press. The particular computer game of Stockfish vs. Spark, discussed in an earlier chapter, was staged by a Norwegian chess enthusiast named Martin Thoresen. Martin ran web tournaments that paired the world's best chess-playing computers against each other, concluding with an "elite" tournament that paired off the previous winners. Martin wanted to find sponsors to help him cover the electricity bills but no one was interested, and in the spring of 2011 Martin stopped running his tournaments.

Hardly anyone noticed. A few people protested in the comments section of his blog, such as the one fellow who expressed the superiority of chess over time spent with women. (As of early 2013 some version of these matches seems to be restarting; we'll see.)

One of Martin's last tournaments went under the glamorous name of TCEC S3 Stage 2a.

At the same time as TCEC S3 Stage 2a was being run, the so-called Candidates Matches, among human beings, were being played in London. Those matches determined who played Vishy Anand for the world (human) chess championship. Interest in those qualifying matches was slack—by usual terms—because fan favorite Magnus Carlsen, the Norwegian chess prodigy, decided not to play. That's a bit like when Michael Jordan took a few years off from basketball and interest in the NBA Finals fell. Still, unlike with the higher-quality computer matches, many thousands of people followed the human contests and watched Boris Gelfand emerge as the challenger to Anand and eventually lose the championship match.

Despite the humans on the teams, even Freestyle chess isn't very popular, not even by the standards of the chess world. Some chess traditionalists are uncomfortable with the fact that some top Freestyle team members aren't very good at traditional chess, but I don't think that is the main problem. The anonymity and shifting names of the teams make it harder for fans to follow the contests and identify with favorites. It is more difficult to construct narratives about the players, their career arcs, and their personalities and emotional struggles. It doesn't fit the standard model of heroic achievement and struggle against adversity.

Limited spectator interest has meant that Freestyle tournaments have been online only, which makes it harder to attract spectators and corporate sponsorship. There is one wealthy patron operating from United Arab Emirates, but so far Freestyle chess does not seem to have a financial future. Nelson Hernandez estimates that during the key Freestyle events from a few years ago about one hundred teams would play and about one hundred "spectators" would follow the games online. That's hardly setting the world on fire, yet these are among the highest quality and most spectacular chess games ever played.

Might a lot of fans prefer seeing games with human—all-too-human—blunders? It's fun to second-guess the champions and see them humbled by their foibles and their weaknesses. It seems we care more about drama than about perfection. Do you remember the opening to that old TV show *ABC's Wide World of Sports*—"the thrill of victory, the agony of defeat"? It showed a skier taking a violent and dangerous spill, and something pretty similar happens to humans, but not computers, on the chessboard. We like to watch. Insofar as people enjoy watching computers play chess, they enjoy watching the computers indicate when the human contestants are making a mistake.

We do and experience things differently when we feel we are dealing with a machine rather than a person. I have heard Ken Jennings speak of his epic *Jeopardy!* contest with Watson, and how he regrets trying unusual strategies to "psych out" the machine. It didn't work. A similar critique—that he let the mechanical nature of his opponent "get to him"—has been leveled against Garry Kasparov's play in his loss to IBM's Deep Blue in 1997. Unlike with his human opponents, there was no hope of psyching out the computer or probing for its emotional weaknesses over the board. The early days of the Mechanical Turk chess-playing machine show similar flaws in the human opponents of the machine. One historian of the device wrote: "Some players seemed to begin their games with too much confidence, realizing too late the great strength of their clockwork adversary. Others tried bizarre moves, hoping perhaps to confuse the machine, but such efforts nearly always failed."

What does that all say about us as spectators and human beings?

We don't want to treat human beings and computer programs the same way, even when the latter become extremely skilled, or perhaps especially when the latter become extremely skilled. We wish genius machines to serve our practical ends, but we don't want to turn over to them the spheres of life that structure our narratives, drive our emotions, define what our lives are all about, and help us separate right from wrong. We're determined to "keep them in their place."

That's understandable, but it also shows we are a bit intolerant of alien intelligences. It shows we will remain reluctant to consult the wisdom of machine intelligence when it comes to our personal lives, such as our romantic decisions or whether we should take all of our medications. We won't always listen to business or negotiating advice from the genius machines, and maybe we won't be as interested in the music they compose, buying it only if a human composer pretends to have been the creator or co-creator.

For better or worse, we will remain—at the margin—rather desperately in need of help from the genius machines.

PART III

The New World
of Work

9

The New Geography

Looking back, we have seen a great stagnation of wages in the United States since about 1973. Given that this book offers a view of how that era of stagnation is going to evolve into a new chapter in our nation's history, it is worth addressing how much of the stagnant wage trend in the United States was or might remain due to foreign competition. Should we blame the foreigners for our difficulties? And how will foreign competition shape jobs and wages going forward?

Many economists are skeptical of arguments that lay the blame for our weak job market on foreign trade. The notion that foreign competition causes low wages and unemployment has been around for centuries. Yet foreign competition continues to grow, and for the most part, at least until recently, wages have continued to rise. Furthermore, there are plenty of highly open, high-trade economies with employment success stories, most notably Switzerland, which at the end of 2011 had an unemployment rate of only 3.1 percent. Sweden's story is broadly similar.

Economists have been investigating the claim that foreign competition destroys jobs for a long time. It remains difficult to substantiate that claim. It is easy to throw around charges that American workers now have to compete with billions of new workers, many from formerly Communist or Socialist countries, yet most of those billions are not serious competitors, most of all because they have very low productivity. We see also that many service jobs, from filing clerk to cashier, don't face a lot of pressure from outsourcing. Yet the wages in those sectors, over the last few decades, have hardly blossomed. That suggests outsourcing is not the main culprit, and furthermore, technology and machines have historically had a larger effect on labor markets in the United States than has foreign competition. The most detailed study of labor's falling share in output finds that new information and communications technologies—which can substitute for labor—play a larger role in compensation shifts than does foreign trade.

It's also hard to find serious evidence that immigration has hurt American wages in a significant way. Harvard professor George Borjas, a leading critic of our current immigration policies, has presented evidence that immigrants have lowered the wages of high school dropouts, in the long run, by 4.8 percent. But the wages of many other Americans have risen, and some major groups, such as the college educated, have suffered a long-run loss of 0.5 percent in wages, which is close to no effect at all. And that's what the major immigration *critic* finds. Other estimates of the effects of immigration are considerably more positive in terms of the effect on American wages. Papers by Giovanni Peri, among many others suggesting that immigration boosts real wages for most American workers, look at the wages of immigrants across different American cities. The low wages show up in the places that don't attract many immigrants, such as the industrial heartland in the Midwest; a lot of the biggest wage gains come in places that do attract a lot of immigrants, such as the coasts. (This positive correlation between immigration and wages seems to hold up even if we adjust for confusing factors such as the fact that the more rapidly growing cities will attract a greater number of immigrants.) When it comes to wage stagnation, immigration is at most a minor contributing factor.

Outsourcing and Immigration

So workers coming to our country is not a situation that needs to be limited. What about exporting work to workers outside the United States? Some of my economist friends will hate this: It is increasingly hard to deny that outsourcing is playing some role in stagnant American wages and slow job creation.

It's simple. Hiring someone is an investment. If some jobs are becoming "higher investment value" while others are becoming "lower investment value," entrepreneurs and their companies will put the lower investment value jobs in cheaper, lower-wage countries. Some of those jobs will stay in the United States, but only by paying lower wages than would otherwise be the case.

Economists use the forbidding phrase "factor price equalization," which means that if an apple sells for two dollars in the United States and one dollar in Bolivia, there is an incentive to ship apples until the prices move closer together. When presented in terms of the market for apples, that's not very controversial. Yet a similar mechanism operates for labor: If workers in India or China are much cheaper, and not correspondingly unproductive, either the workers will move to the capital or the capital will move to the workers. For the most part it has been the latter—it is easier to move the capital to the workers than the workers to the capital, if only because of US immigration restrictions. When China, India, and other countries started doing better economically, all of a sudden investors figured out that their workers were undervalued in the global marketplace.

I am struck by one anecdotal, stand-alone fact in particular: Between 2000 and 2009, multinationals cut 2.9 million jobs in the United States and added 2.4 million jobs overseas. That doesn't prove a causal connection, but it would be surprising if those were two completely unrelated numbers.

More systematically, a recent study by David H. Autor, David Dorn, and Gordon H. Hanson supports the notion that outsourcing has had a negative impact on US wages. From 1991 to 2007, US imports from China rose from $26 billion to about $330 billion. They find that greater exposure of a region to Chinese imports predicts weak performance in manufacturing employment, weak wage performance, and, in those same regions, a growing demand for transfer payments ("handouts") from the government.

Another recent study, by Runjuan Liu and Daniel Trefler, finds further evidence of job market problems from outsourcing. They found that "downward occupational switching," meaning people took lower-paid jobs in less lucrative sectors, went up by 17 percent among affected groups, transitions to unemployment increased by 0.9 percentage points, and the earnings of affected "occupational stayers" fell by 2.3 percent.

You shouldn't, from those numbers alone, jump to the conclusion that outsourcing is bad, all things considered. These studies do not pretend to isolate every effect or every possible side benefit of outsourcing. They're asking whether outsourcing has made some segments of the US labor market more sluggish, and the answer is a clear yes.

In the standard economic approach to international trade, those losses in the labor market are offset by greater gains elsewhere. So maybe wages are falling in some sectors but jobs and wage gains in other sectors make up for that. Yet it's hard to find those offsetting gains in the numbers, except perhaps in the returns to capital. The reemployment of displaced workers has been relatively weak since the late 1990s for reasons that are no fault of the Chinese. The rising demands have been for very special kinds of skilled workers rather than for workers as a whole. To some extent the losses in the job market are made up by lower prices for outsourced goods for US consumers, but we don't know by how much.

There's more evidence for the importance of outsourcing, albeit of an indirect sort. Economist Michael Mandel has taken a close look at US productivity numbers, especially for the 2000–2010 decade. In the earlier part of those years, the measured productivity gains in the US economy are fairly high, but real wages are stagnant and there is no new net job creation in the United States. That's a puzzle, since higher productivity usually translates into higher wages fairly quickly. Mandel has done some careful investigations of the numbers, and his hypothesis is that we are confusing two very different kinds of productivities: productivity by making American workers more effective and productivity by outsourcing.

Let's say that an American investor outsources some of his production from Ohio to Shanghai; Shanghai of course has lower wages than does Ohio. The way our productivity numbers are calculated in practice, that job shift will usually show up as a productivity gain. And it *is* a productivity gain, to the owners of the factory and also to consumers, who eventually will pay lower prices for their products. It is *not* a productivity gain for American workers, especially not in Ohio. It's not the same as making American workers more productive by giving them better machines, better training, or a neat new product to make.

Mandel has come up with circumstantial evidence that a lot of our recent measured productivity gains are of this sort. For instance, a lot of the sectors where outsourcing is prominent are the sectors with very high productivity gains, such as producing iPads. The flow of capital, and the introduction of further foreign competition, leads to downward pressure on American wages or at the very least it prevents American wages from rising more rapidly. I wouldn't say that Mandel has produced a smoking gun, but mismeasured outsourcing is an obvious missing piece that would make the macroeconomic and productivity numbers fit together. (Another piece of the puzzle may be what I called in chapter three "restructuring productivity" gains: firing workers who weren't producing much value instead of raising the productivity of the workers who remain.)

Economists don't have problems with the idea that free trade and investment have not, in recent times, brought huge gains to the economy of central and northern Ohio. I'm asking them to take that perspective one step further and apply it to a larger part of our nation.

Still, it would be a mistake to think foreign competition, or additional outsourcing, has been the fundamental problem with the American economy. First, outsourcing leads to lower prices for consumer goods and that helps maintain or boost real wages to some extent. iPads are pretty cheap, in part because they are made with low wages. iPads also will, in the longer run, boost our healthcare and education sectors. (And besides: They're just plain fun.)

Outsourcing is still a smaller phenomenon than many people think. Currently imports from China are measured at about 2.7 percent of US consumer spending. Furthermore, for each dollar of imports from China, about fifty-five cents was ultimately spent in the United States preparing the import in some manner. That leaves Chinese imports, measured in terms of true net impact, at about 1.3 percent of American spending. Admittedly, that spending is concentrated in higher percentages in some particular sectors, such as toys and final assembly of electronic products. It's reasonable to see the Chinese production, and investment in China, as limiting investment in American jobs, but outsourcing just isn't big enough to be *driving* the problem of stagnant wages. It's more like one of several reasons why a lot of low-wage American workers aren't re-employed as rapidly as in times past.

It is too easy to press the "now we are competing with 2.5 billion Indians and Chinese" argument too far. Americans have had competitors for a long time. For instance, the 1948–1973 period is often considered the glory years of industrial America. Yet during this time the Western European nations rose from wartime destruction to near parity in living standards with the United States. Japan grew rapidly too. Those same nations competed head-on with the United States in a lot of major industrial sectors, including automobile manufacturing and electronics. This competition did cause problems for particular sectors of the American economy, but overall the United States did very well during those years. The period 1880–1929 saw phenomenal growth in what we now call the developing world, including Latin America. Many countries moved from being a series of isolated rural outposts to having real cities and real manufacturing sectors and real connections to the global economy; Argentina is perhaps the best known example but the same is true for Mexico. The American economy did fine during that period too, even though it was an intense period of globalization and foreign competition, including the outsourcing of at least some US manufacturing.

During these periods of prosperity we were world leaders in education—K–12 and university. There was a closer match between the skills required of workers at higher levels of the value chain and the skills that American workers actually possessed. Nowadays, the demands of machinery—including of course computers—are rising at a faster rate than are human capabilities. The machines are getting better education, more rapidly and more cheaply, than are their human teammates and potential teammates. That's the root of the problem for a lot of workers.

Many millions of people can turn a screw on an assembly line, work a lathe, or handle a telephone switchboard. Not so many people can team up fruitfully with Rybka or, more generally, work with intelligent machines operating in, say, our finance, service, and medical sectors. To blame the Chinese, or outsourcing, is to point to a mere shadow play, to confuse superficial events with the ultimate cause of the slowdown in wage growth.

By the way, the very best US chess players are often foreigners in a sense—most of them having come from Israel and the former Soviet Union. You might think they've "taken slots" from US players, but the deeper reality is that having lots of good players in your country, foreigners or not, helps—not hurts—your chance of becoming world champion. In any case, you need to be able to beat the best, so exposure to great players at a young age is a big help.

When it comes to outsourcing, three truths need to be observed.

First, there is no realistic way to stop Americans from investing abroad. Rather than closing ourselves off to the world, we need to increase our domestic productivity and improve education to create more productive workers, including those who can work Freestyle.

Second, at a fundamental moral level a job for "a foreigner" is every bit as worthy an outcome as a job for "a real American." If Chinese wages are going up and American wages are somewhat flat, as has been the case, I say bravo to them and let's try to meet the challenge. International trade and investment still produce wonderful gains for the global economy as a whole. It may ring hollow with the American electorate, but it is still true.

Third, if you're worried about outsourcing, you should probably have a more liberal rather than a less liberal attitude toward immigration. If the United States takes in more immigrants, the areas in which those immigrants work are less likely to see jobs outsourced abroad. Immigration makes it possible to keep those jobs at home. In fact, the bigger a threat outsourcing becomes, the more important immigration is for keeping us competitive and for keeping other complementary jobs in place.

When companies move production offshore, they pull away not only low-wage service jobs but also many related jobs, such as high-skilled managers, tech repair people, and others. But hiring immigrants for low-wage jobs helps keep many kinds of support services in the United States. In fact, when immigration is rising as a share of employment in an economic sector, offshoring tends to be falling, and vice versa. That means immigrants are very often competing more with offshored workers than with other laborers in America.

Here is another way to think about the problem. A lot of

economic activity wants to locate itself in those regions that are the most populous, wealthiest, most important, and most central to the nerve center of our world. Not everywhere can be a center of the global economy and not many people and investors wish to focus their activities in Nebraska. China has a lot of people and a lot of activity and to some extent it is taking some of this attention from the West, including from the United States. Part of a competitive solution, for the United States, is to build up its own economic and cultural clusters of importance. That means more people and also more immigration, both high-skill and low-skill.

Is the dominant economic cluster going to be the northern plains of China, combined with a few of its southern port cities? Or will it be the two coasts of North America, with a boost from the heartland for agriculture and fossil fuels? Or will there be two or maybe three dominant clusters, including northwestern Europe?

Some of what is going on in today's global economy is a reorienting of economic activity toward where most of the people are, and obviously, most people live in Asia. There have been a lot of people in Asia for a long time, but these days they have a much higher per capita income. One way to compete, to remain a global center of economic activity, is to build up a larger and stronger cluster of population and production. If we want North America to be a global leader, we should bring in more people to the United States. We don't want the United States, and the liberal values it (sometimes) projects, to become a second-rate force on a global scale, neither economically nor politically.

Immigration is vital to the future economic vitality of the United States. If that immigration is Latino, as indeed it often is in the United States, the longer-run effect is to build up entrepreneurship and democratic values in the other countries in this hemisphere.

Over the last twenty years, most of Latin America has become more prosperous and it also has become either more democratic or more sustainably democratic, with Cuba and Venezuela as two notable exceptions. It is hard to say how much influence the United States has had on those processes, but we should be trying to have democratic influence, and that means allowing a large number of Latinos to live and work here, to earn some money, and to absorb democratic and broadly liberal values.

A Global Geographic Trend

Our cluster isn't just the major cities of the United States; in the bigger picture our cluster is this entire hemisphere, including Canada and Mexico. China's cluster includes significant parts of Asia, and we already can see parts of Southeast Asia evolving to become economic satellites of China. The point is not to fight an economic war between "us" and "them," but we do want our cluster to attract investment and interest, and that means the United States has a significant stake in the fortunes of Latin America. Shutting it out isn't the way to go, but rather we should try to work with these other nations. Latino immigration brings business contacts, joint linguistic facility, and a general sense of sharing a somewhat common destiny. We North Americans should be very happy that most of Latin America has been doing very well over the last decade.

Just as labor market outcomes will move toward the poles of either "very good" or "very bad," so will the same be true for a lot of cities, states, geographic regions, and countries.

What we see happening is that individuals with college degrees are gravitating to areas where a relatively high percentage of the other individuals also have college degrees. Some of the winning areas are Raleigh, North Carolina, San Francisco, and Stamford, Connecticut, where over 40 percent of the adult residents have college degrees. You can add select areas of New York, Chicago, and Los Angeles to this list, although those cities as a whole do not show uniform progress in recruiting educated individuals. Some of the loser cities include Bakersfield, California, and Youngstown, Ohio, where the percentage of educated adults is less than one-fifth. It should come as no surprise that the cities with high levels of education tend to have much lower levels of unemployment. We see also that in terms of per capita income, the poorer regions of the United States are no longer catching up to the wealthier regions.

This is a big change from the America of times past. Circa 1970, the most educated and least educated cities differed from each other by about sixteen percentage points, and most metropolitan areas were within five percentage points of the average. Today the difference between the most and least educated cities has about doubled in terms of percentage points, and only about half of US cities are within five percentage points of the average level of education. Ambitious and talented young people today are more likely to want to live in a relatively small number of cities and regions, rather than spreading themselves out as much as they used to. That is why Brooklyn, one relatively affordable part of New York City, is becoming a center for tech innovation.

It's also interesting to look at the geographic concentration of the wage benefits that result from the internet. Investment in the internet correlates with wage and employment growth in American counties that represent about 42 percent of the American population. In the other locales wages have not benefited from the internet at all, and so it can be said that the internet has increased regional inequality in the United States. This is far from the utopian dreams of the early days of the information economy.

It is worth considering a little more exactly the new ways in which distance does and does not matter. Because of the internet and Amazon, among other developments, it is easier to become self-educated in many more different parts of the world. It is also easier to have a "good enough" or low budget (but happy) life in many more different parts of the world, again because of technology. But if you wish to be a high earner, learning from other well-educated people, geographic proximity is growing in importance, whether in companies or in leading amenities-rich cities or most likely in both.

We are seeing similar tendencies in Europe. For some time now, London and much of south England has been quite wealthy, while most of northern England has done less well. If you are born in England, or if you wish to move to England, and you have ambition, you will probably prefer to work in London or in the south not too far from London. It is less clear that you will prefer to retire in London, and in fact the very commercial success of London makes retirement there more costly and thus more problematic. Furthermore, it is not obvious that your children will need to grow up in London to stand a chance of becoming math prodigies.

We can expect to see a similar process of geographic segregation for the Continent, although it will take much longer for reasons of language barriers and legal difficulties in moving workers across borders. Southern Germany is an extremely productive region and we can expect it will become more crowded and also more productive. The top German firms are generating some impressive learning curves. If you are a top Spanish engineer, you shouldn't take it for granted that you ought to stay in Spain. Quite possibly your productivity will be much higher in Stuttgart or Munich. Indeed, we already see Spanish engineers moving to the wealthier parts of Europe, most of all Germany, and it increasingly looks as if many of them will end up staying there. Suddenly the German population is growing again, and that is because of migration into Germany, not the birth rate from native German citizens.

I hardly expect Spain to be emptied out, but soon observers will start to realize that "economic integration" isn't exactly working out as advertised. What people expected from economic integration was a wealthier and slightly more ethnically diverse version of what they had ten or fifteen years ago. What they will be getting is a dramatic shift of labor resources into the most highly valued firms and also into the most highly valued business regions. There will be lots of "hollowing out" of various regions—at least in terms of well-educated high earners—and not everyone will be on the winning side of this process. More areas will specialize in being locations for tourists, retirees, and individuals living off family assistance or government assistance.

This point suggests a broader perspective on the economic crisis in the Eurozone. Most current discussions of the crisis focus on debts, failing banks, capital-flight deflationary pressures, self-defeating austerity, and so on. That's all relevant, but there is a deeper backdrop to these problems. Europe really is integrating economically, however imperfectly, and we are learning that there are big winners and big losers to this process. The debt problems and the high bond yields and the other signs of crisis can be read—in this account—as our financial markets heralding the future arrival of major changes in economic geography. A lot more of Mediterranean Europe is going to look like southern Italy and Sicily: somewhat empty in terms of economically successful enterprises on a large scale. It will rely more on tourism and more on retirees and government transfers, but overall the fiscal problems of these regions will worsen and they may lose some political autonomy. In any case, their current and forthcoming revenue problems will show up in forward-looking market prices, in capital markets most of all, and that means crises in the bond markets as indeed we have been seeing.

It is striking how frequently the phrase "the periphery" is used in discussions of the Eurozone. It is exactly appropriate, and few people hesitate to use it to describe large parts of a region devoted to the ideal of economic *integration*. It's time we started thinking through the other implications of that reality, namely that the geographic and institutional clustering of commercial talent is on the rise.

The Good of Foreigners

Of course, public opinion isn't usually rational, so most people do not look to immigration to help fix the offshoring problem. A lot of people are "anti-foreigner" and this makes them skeptical about both offshoring and immigration, because both phenomena have something to do with foreigners. People get into a broadly "anti-foreigner mood" and then they apply those feelings somewhat indiscriminately. The more subtle reality is that immigration gives us some protection against offshoring and probably helps keep jobs in the United States. It also helps protect the long-term position of the United States as the world's most dominant nation, whether economically or politically or culturally.

I don't fear a big surge in protectionism in this country, or in the world more generally. A recent study by Michael Spence, a Nobel laureate in economics, and Sandile Hlatshwayo shows that almost all of our recent job gains (the gross gains in some sectors) have come in so-called non-tradable sectors, such as health care and government. People who work in government, health care, and education just aren't that worried about foreign competitors or even outsourcing.

That security of these non-tradable sectors is nice for many of us, but it also means that people in most newly created jobs in the United States aren't facing so much of a daily market test. Most of our job growth is coming in what I call low-accountability sectors. People get paid to produce things, or offer services, and we're never quite sure how much value they are putting on the table. The value they produce, or the lack thereof, is never subject to much of a market test. Of course, that may not augur well for our ability to upgrade our future productivity, even if today a lot of these workers are producing fairly valuable goods and services.

One bright side, if you could call it that, is that hardly anyone hates free trade these days, compared to, say, the mood of the 1980s. The jobs that were eliminated by free trade are already gone, as large numbers of manufacturing jobs have moved to lower-wage countries or have been eliminated by the greater productivity of machines.

The good news for Americans is that the longer-term trends seem to favor the relative position of the United States in the global economy. The United States is likely to continue as a leader in applying artificial intelligence and this will likely cement long-term American economic growth.

One reason for this is simply that America can sell its artificial intelligence products to the rest of the world, but a deeper mechanism is operating too. If there aren't many workers in a plant or project, hiring in a wealthy country brings less of a wage penalty. Indeed, some manufacturing plants are moving back to the United States—"reshoring" it is called—and those are the plants that use lots of robots and artificial intelligence. The technical proficiency is in the West for the most part and yet these plants will still create American jobs indirectly—through infrastructure, the service sector, and distribution—even if there are relatively few human jobs within the plant itself. Mexico, with its proximity to the United States and relatively high productivity levels, is also turning into a preferred destination for new manufacturing investments. Canada is helping to supply both the resources and the human talent for a new North American economic renaissance.

When you combine these developments with fracking and the greater production of domestic oil and natural gas resources, the North American nations are well positioned to be dominant export powers going forward. North America also has the logistical abilities, and a strong enough trade-protecting military, to enable its export businesses to prosper. We will likely see a new American Century, or rather a North American Century, and if any countries should be worried it is those with low wages, including China. Unless those countries make the next leap up a technology ladder, competing on the basis of capital as well as labor, many of their current competitive advantages will erode.

10

Relearning Education

We as a nation have been thinking about education without knowing what we really want from it. Do we want well-rounded young adults to emerge? Or good citizens? Role models? These goals seem reasonable but what do they mean? For the purposes of this chapter, and indeed this book, I'll keep the goal simple. One goal of better education is to procure better earnings. How we might achieve that is the question.

Whether we will remain a middle class society or not depends firstly on how many people will prove to be effective working with intelligent machines. One percent of the population? Ten percent? Fifty percent? Secondly it depends on how many people will find work either as personal servants or as more distant service providers to the high earners, and what wages they will be able to negotiate.

The answers will have a lot to do with education. In particular, how many people will receive world-class or at least satisfactory educations in the years to come so they can earn the best, or at least acceptable, wages?

To get a grasp on this issue, it's not enough to look at the workplace competition between man and machine intelligence. Smart machines also produce new and valuable goods and services, and that boosts real wages, even if we don't observe the boss handing out a direct raise. Furthermore, those goods and services may help workers earn more in the future, especially when the services take the form of education. So let's take a look at how education will change, keeping in mind these two blades of the scissors, namely that machine intelligence can replace human labor and augment the value of human labor for many individuals.

Online education is one place where the new information technologies are emerging. For instance, millions of people are taking MOOCs (massive open online courses) or using the free instructional videos from Khan Academy on mathematics and other topics. Circa 2013, no one is surprised when a new foreign aid program consists simply of dropping iPads into rural Ethiopia and letting children figure out how to work them.

Online education is expanding beyond its niche status, but sometimes we don't recognize the most important developments as explicit education. In my own field of economics, what is the most common and regular form of contact the general public has with economic reasoning? It's no longer the Econ 101 class but rather it is economics blogs, which are read by hundreds of thousands of people every day. I submit that "cross-blog dialogue," as I call it, is for many people a better way of learning than boring lectures, Power-Points, and dry, overly homogenized, designed-not-to-offend-anybody textbooks. Schools are supposed to be proper and politically correct, but sometimes the point really sticks when Paul

Krugman calls someone an idiot on his popular blog and explains why—whether or not you agree with Krugman or the (supposed) idiot. Blogs have to get people to care because it is a very competitive environment. The competition is to capture anyone's attention.

It's not just formal online education and blogs. Apps, TED lectures on YouTube, Twitter, reading Wikipedia, or just learning how to work and set up your iPad are all manifestations of this new world of competitive education, based on interaction with machine intelligence. These new methods of learning are all based on the principles of time-shifting (watch and listen when you want), user control, direct feedback, the construction of online communities, and the packaging of information into much smaller bits than the traditional lecture or textbook chapter. A lot of us find these features to be a very appealing blend of qualities and they are why online education is taking off in a way that watching lectures on videocassette never did.

Online education is even growing as a supplement to K–12 or in some cases as a replacement altogether. As of late 2011, about 250,000 K–12 students are enrolled in full-time virtual schools. Over two million K–12 students take at least one class online.

At these online schools, the degree of contact with flesh-and-blood teachers varies. Instructors might answer questions by email, phone, or videoconference, supplemented by periodic meetings, class trips, and "live," in-the-classroom exams. It's often for less than half the price of a traditional K–12 schooling experience.

The world still awaits systematic, rigorous (randomized control trial) studies of all of these methods of learning, and it is too early to say what is working and what is not. Nonetheless, we do know two things for sure. First, very often the online methods are much cheaper and also more flexible than the previous alternatives.

Second, some learners—quite possibly a minority—love the online methods. We can thus expect that online education in its various manifestations will likely represent a fair-sized chunk of the future of the sector. It will remain the case that much of the online product will not be as good as full-time, quality, face-to-face instruction, but many a market battle has been won on the grounds of price. It's already the case that a lot of students don't get a really good face-to-face educational product. Online K–12 education has a future for many of the same reasons that McDonald's has been a successful corporation, even though there can be little doubt that a Big Mac is not nearly as good as the finest dish in the best restaurant in town.

Let's consider a few features of the economics of online education that are frequently overlooked, with an eye toward seeing why it will really matter.

The first is that online education will be extremely cheap. Once an online course is created, additional students can be handled at relatively low cost, often close to zero cost. (We can even look forward to the day where essay questions are graded by artificial intelligence, and indeed some examples of this already are succeeding.) Over time, competitive pressures will operate to push price down close to costs. That's not quite the world we have today, because setup costs for the classes remain a burden and most good colleges and universities have radically incomplete online offerings. It is also true that getting official accreditation for these courses is far from easy.

Once online materials are up and running, and properly tested for effectiveness, the next step is for some schools to expand the sales of their online materials to other schools, and that is when we will start to see very low prices indeed, driven by competition. It will be a brutal age of good schools and also mediocre schools undercutting each other in terms of price and thus tuition revenue. If it costs $200 to serve a class to another student, how long will it be before an educational institution undercuts a competitor charging $2,000 for those credits? I don't think the price will fall all the way to $200, because good schools won't want to look too cheap, and maybe they don't need the money, but still I expect the price for a class to be much lower than its current level, especially at institutions below the top tier. Of course the most successful classes will make up the price discount with volume in a diverse international English-language market for education.

Second, online education is more flexible than many people believe. Many think that students won't pay attention to a computer or a machine teaching a lesson, but there are many ways of using the online product to lower costs while keeping the attention of the students. Imagine keeping students on campus and having them show up in classrooms at regular points in time, as they do now. Or let them show up when they want. Put a large number of those students in a large "barn" and have them watch the online lessons, with a few (non-tenured) instructors wandering the rows to help out those with problems.

Does that sound Orwellian? Virginia Tech is doing that right now for some of its math classes and it seems to be working out just fine. It is called the Emporium model and it is in use in at least one hundred schools. In the case of Virginia Tech, it covers classes from 200 to 2,000 students, it has 537 computers as of last count, it is open 24/7, and they converted a former department store to this use; the space is set in a shopping mall with Muzak and the other stores as a draw. Imagine adding to this mix enhanced capabilities for machine intelligence to monitor and report which students aren't working through the material at all or at the right pace. The costs for the instructors are of course much lower.

This Emporium model may or may not be the future of the sector or it may just fill one particular niche. The point is that most of the off-the-cuff critiques of online education ("no one would do *that*!") underestimate how versatile the idea can be and how much it can be combined with face-to-face techniques in a hybrid manner, at least when necessary.

Third, another big change is that the profits from developing teaching innovations will be much higher than they are today. Let's say that today, or rather in the pre-online world, I discovered a much better way to teach some economic principle, for example the idea of opportunity cost (how doing one thing means you can't do something else). I might try to write a successful textbook but that is a hard market to crack and it requires presenting lots and lots of ideas in a good way. Without the prospect for a textbook serving a mass market, I am innovating for a relatively small number of students and maybe a pat on the back when it comes time for teaching evaluations. But when the best courses serve tens of thousands or hundreds of thousands of students, or maybe even millions, the financial returns to pedagogical innovation will be looked at in a new light. Imagine writing "the opportunity cost app" and having it incorporated in economics instruction around the world. As a society, we'll put a lot more effort into teaching things better. For all the virtues of human, face-to-face instruction, it fares pretty miserably when it comes to economies of scale and scope.

Fourth, online education also allows for a much more precise measurement of learning. Consider the Khan Academy and its online videos. They are already measuring which videos lead to the best performance on quiz scores, which videos have to be watched more than once, at which point in the videos individuals stop for pause and replay, and so on. We are creating a treasure trove of information about actual learning, and we are just beginning to mine this data.

If a student is falling behind, or in denial about his or her progress in the course, the software is the first to know. We're about to apply "Big Data" to the students themselves, and man and machine will work together to improve significantly the quality of education. In a slightly more distant future, we can imagine the computers hooked up to bodily sensors of pulse and scans of facial movements, perhaps to determine if the student is bored, distracted, or simply not understanding the material.

Learning Games

Computers have taught many people, young and old, to play better chess. It can even be said computers have revolutionized the way the game is taught.

Beginners through to intermediate players can learn the moves, tactics, and strategies, and how to play the opening and endgame. The programs offer preset positions, more rapid drills, categorized problems, and embedded grading, for either a formal lesson from a teacher or for self-instruction. For younger children (and sometimes adults too) the programs offer attractive video graphics, accompanying characters (*Fritz and Chesster* is one popular game), and sometimes aggressive crunching sounds when the pieces are captured. Chess instruction draws upon the skill of the programs but it also uses advances from video and computer games. Some of the programs now can ask essay questions about why a particular move might be a good one, and they can grade the answer.

The most important teaching function, however, is that a player has a willing partner and analyst 24/7. The programs can analyze chess games on request, including one's own games. Just enter the game into the program and request feedback in the form of suggested improvements and analysis. What's interesting is that this function is used more or less the same way by both grandmasters and beginners, albeit with different levels of understanding. In other words, the younger and lesser players can move directly to the best learning strategies employed by the experts.

There is plenty of talk about finding good role models for how to do education, but not enough people are focusing on the games industry, including chess but by no means restricted to it. It's remarkable what a good job the games industry, especially in its online manifestations, has done educating us in playing the games. Of course there is self-interest there; game manufacturers are selling us more games as a result. But the educational accomplishment and intellectual advances are remarkable. Many of those role-playing, time-extended, multiplayer games are forbiddingly hard or at least so it seems. They involve hundreds or thousands of people manipulating hundreds or thousands of virtual characters, not always human, spread across (virtual) geographic space and engaged in trade, battles, elections, and many other activities, all governed by complex rules within the game and governed by complex software at the level of player interface. And yet people learn those games nonetheless and often master them.

The many millions of people these games have educated include millions who would not otherwise be thought of as educational winners or job market winners. It is actually the most astonishing education success story of our time and it is driven by commercial incentives and the desire to make learning fun. Game-based interaction tends to be hands-on and step-by-step, moving up a ladder of complexity with rewards along the way to keep the game player interested. Of course these games don't interest everyone, but wouldn't it be funny if we already had figured out the education problem and simply didn't know it?

If there is anyone today who understands the dynamic potential of games it is Jane McGonigal (game designer and author of *Reality Is Broken: Why Games Make Us Better and How They Can Change the World*). Her dream, entirely reasonable in my view, is to see a games designer nominated for a Nobel Peace Prize.

For all the successes of games, however, they also point out some limitations of education by computer, at least how we currently practice it. Education into the world of games works remarkably well, but it works mainly for people who wish to learn the games. Chess-playing computers don't boost the play of diffident students who refuse to spend much time with the machine. For people who aren't already motivated, or on the verge of picking up a new fascination, nothing about the game is all that enticing or seductive.

Sometimes a student may care about doing well with grades but not about mastering the actual material and moving on to the next step. Chess teacher Peter Snow reports that some of his young students love playing against the computer, but they deliberately put the quality settings on the program so low that they can beat it many times in a row. At this point they should raise the skill level of the program to make the challenge tougher, but they don't always want to do so. Similarly, studies of spelling bees show that the winning spellers are those who not only work hard, but who engage in disciplined forms of study that do not always yield immediate positive feedback.

We see similar issues with online courses and videos: They are great for those who are already inclined to learn more and who take an interest in the "game" of mastering this material, be it in astronomy, modern poetry, or quadratic equations.

New Higher Education Models

If computer games are so good at rapid education, especially with young people, where are the results? They are most apparent at the higher levels of achievement. For instance, the average age of chess prodigies, like most game prodigies, has been falling over time, and a large part of this is driven by better mechanical education.

In the 1950s, it was considered miraculous when Bobby Fischer became a master at age thirteen. David Pruess, who writes for Chess.com, acutely observed, "Today, you have 7- and 8-year-olds who are training better than Bobby Fischer did a generation before." A typical anecdote is from Jesse Kraai, then a twenty-eight-year-old grandmaster. To win a tournament in Reno, Kraai played four of his six matches against children, with an average age of thirteen. This is historically exceptional. And it isn't only happening in chess. The worlds of programming and social networks have become areas where individuals can make truly fundamental contributions at quite young ages, as did Mark Zuckerberg with Facebook. More generally, some recent research by Jonathan Wai, Martha Putallaz, and Matthew C. Makel on prodigies does indicate that indeed the smart are getting smarter and at younger ages too.

The superstars will reach higher and more dramatic peaks, and at earlier ages. Magnus Carlsen is, as I write, the highest rated player in the world and arguably the most impressive chess prodigy of all time, having attained grandmaster status at thirteen and world number one status at age nineteen, the latter a record. He is from Tønsberg, in southern Norway, and prior to the computer age Norway has no record of producing top chess players at all. Even Oslo (Carlsen now lives on its outskirts) is a relatively small metropolitan area of fewer than 1.5 million people. Carlsen, of course, had the chance to play chess over the internet.

Many more young chess players come from the far reaches of the globe, including distant parts of China and India. The top Chinese and Indian players grew up playing against computers and learning from computers and playing online.

In the old days the Chess Olympiad was dominated by the Soviet Union. It wasn't just the state subsidies for chess and the restrictions on using one's mind for business or artistic creativity (a lot of talented people found chess to be their best or perhaps only option). Most other parts of the world had genuinely backward chess scenes. But now you can put yourself on the path to top skills from almost anywhere. In 2008 and 2012 the small nation of Armenia took first place at the Chess Olympiad, and unlike in the old days the top Armenian players do not need to move to Moscow. The country, with about three million people, is a perennial contender for top chess honors.

Again, we see some analogous results popping up in online education. When Sebastian Thrun, then of Stanford, taught his artificial intelligence course online, the best performers were not the students from Stanford. Generally the best performers were the students abroad, often from poor countries and very often from India. All of a sudden these individuals had a chance to outperform the

US domestic elites. It is no surprise that recent speculation has centered on whether tech employers and other companies might use online courses as a new way to recruit talent.

Online education can thus be extremely egalitarian, but it is egalitarian in a funny way. It can catapult the smart, motivated, but nonelite individuals over the members of elite communities. It does not, however, push the uninterested student to the head of the pack. Here is yet another way in which the idea of a hyper-meritocracy will apply to our future.

The lessons are clear: Workable machine intelligence means that a good education no longer relies on living near a major city. Hanging out with the elites isn't as important as it used to be, unless you have some outlandish idea of the programs themselves as the new elites. The fact is that when humans and computers work together and cooperate, the rewards flow more readily to top talent, not to the socially well connected. Machine intelligence is the friend of the educational parvenu, albeit the disciplined, gutsy parvenu with high IQ.

In terms of absolute levels of performance, machine-driven education will boost strivers at many different levels. In terms of relative performance, it is actually many of the non-top performers who will rise the most, because many of the very top performers would attract a lot of attention and instruction under any system, with or without computers. And then many individuals will not rise at all—they won't sit down at the screen.

It's amazing how quickly the computer revolution has spread to chess and chess teaching. Readily available, affordable chess programs arrived only in the late 1990s. Since that time, computer programs have taken over the teaching of chess and have changed how the top players improve their game at the highest levels.

It remains to be seen whether online education will spread with equal rapidity but most likely it will not. One major problem is simply that universities are for the most part bureaucracies. Faculty often fear online education because they sense it will either put them out of a job, lower their status and importance, or force them to learn fundamentally new methods of teaching, none of which sound like pleasant prospects, especially for a class of individuals used to holding protected jobs that involve a certain amount of autonomy and indeed coddling.

And the faculty are by no means the only obstacle. One central question is how quickly accrediting bodies will move to grant full and transferable credits for good online courses. I do expect to see some progress in this direction, but accreditors serve in part to prop up a higher education cartel. At some point they might think twice about allowing so much competitiveness into the market, knowing that it will knock away lots of revenue from the institutions—colleges and universities—that pay their fees. I do suspect the genie is already out of the proverbial bottle, but battles over accreditation will still have to be fought. Note that Sebastian Thrun, after a successful experience with teaching online material, decided to quit his tenured teaching position at Stanford and start his own educational company.

When it comes to chess, no special interests interfered with the transition to machine-based instruction. No accrediting bodies asked these search engines to prove themselves. There were no lesser, "tenured" computers that had to be pushed out of the way, edged into retirement, or waited out until they passed away. Competition did the job. The transition took only a few years and it met with little opposition. It's one of the world's most startling educational stories and yet it is not a part of our national discussion about education.

When it comes to computers and chess instruction, the computers and programs are not just an "add-on." They are not merely something an instructor pulls out of his pocket to make the lesson more exciting or more exotic (although that happens too). The chess programs are the very center of the teaching, and for the best students they end up being the most important teachers themselves. The computer becomes the center and—dare I say it?—the human teacher becomes the add-on, however important an add-on that may be.

Again, you can see why that might not prove so popular at the faculty club.

We see also a growing role of machine-aided self-education when it comes to the iPad and many other computer-related devices and programs. The kids who learn iPads start by trying to manipulate them and they learn by trial and error, allowing the iPad to teach them. There is no iPad instruction booklet in the box you bring home from the store. You can download one online, or read it online, but hardly anybody does.

This kind of machine-based learning is driven by a hunger for knowledge, not by a desire to show off your talent or to "signal" as we economists say. If you're not a good player, the fact that you studied with a top teacher doesn't mean a thing. No one is impressed and no one will want you to play for their team. There is nothing comparable to the glow resulting from a Harvard degree: Announcing "I studied with Rybka" would bring gales of laughter, since anyone can do that. Indeed, the best chess teacher takes all comers and has every incentive to do so. The company selling Rybka tries to make its product replicable and universal, whereas Harvard tries to make its product as exclusive as possible. Now, which model do you think will spread and gain influence in the long run?

Of course Harvard, MIT, Stanford, and other schools of that ilk may end up being the ones providing the online education. MIT and Stanford have already been pioneers in that area, and given the quality of these institutions and their faculty, it would hardly be surprising for them to continue playing a leading role. Yet there is a fly in the ointment, and it remains to be seen whether schools such as Harvard can excise it. The current business model of Harvard and Princeton is to market the quality of *exclusivity* and to raise money by encouraging alumni to donate to such a wonderful and exclusive institution. Over time the size of their entering classes doesn't go up a whole lot, even though the number of highly talented individuals who apply has gone up a lot. Exclusivity forbids it.

Given that business model, will Harvard and Princeton really be the ones to award *credits* for online courses, say to several thousand very good students in Bangladesh? Imagine a really bright seventeen-year-old in Dhaka, walking around with a big grin on his face because he earned three credits—Harvard credits—from Harvard's intro to economics class. (Those credits would likely be good at higher education institutions in Bangladesh and elsewhere.) I am not sure we are close to that day because I am not sure Harvard will give up its exclusivity so readily. Just think how many "partial Harvard alums" we would have walking around.

But say your school is not Harvard but rather your school is ranked number thirty-two in the United States and it wishes it could somehow steal a march on Harvard. It's not the most prestigious school but it still has a lot of talent and a very good name. Maybe such a school will offer three credits for its online course well before Harvard does. In that world, which service will the people in Bangladesh prefer? A course with Harvard and a pat on the back? Or a course with, say, three credits from NYU? I suspect for the majority of the school-age market it will be the latter.

Face-to-face Instruction

I mentioned at the beginning of this chapter that learning with machines would modify rather than replace face-to-face instruction, so what will the new world of face-to-face teaching look like?

Precisely because of computers, the chess instructor no longer has to spend as much time teaching the student how to analyze positions; the computer does this. The chess instructor is no longer a unique chance for the weaker student to have a crack at playing against a better player; the computer does this too. The computer can also suggest an opening repertoire and perform many other functions that formerly required the chess instructor. In part, the human chess instructor teaches the pupil how to use the computer. The human instructor has also become more important for motivation, psychology, teaching pacing, and teaching the psychological foibles of potential human opponents. With younger and less experienced players, the skills include keeping one's composure, maintaining concentration, and not getting psyched out or intimidated by older or better opponents. These skills are important, and if anything they are more important outside the world of chess.

The next step is that human instructors will consult the machines to better understand the mistakes their students are making. Even if the machines do the work measuring the mistakes, as discussed earlier, the human instructor may still be the one to interpret and deliver that information (in inspiring fashion of course), and the one to outline a course for improvement.

If it's a thirty-year-old instructor of chess or anything else teaching a ten-year-old novice, that teacher is first and foremost a role model and a motivator and to some extent an entertainer. He is a flesh-and-blood exemplar. He shows that success is possible. He exudes enthusiasm for the pursuit of knowledge, or he is not going to make a very good teacher, no matter what his level of expertise.

For all these reasons, chess lessons on Skype, as you might commission from India, have not become popular, even though they are cheaper than face-to-face instruction. The programs have forced chess instruction to evolve, in largely beneficial ways, and—here is a key point—in ways that make the job harder to outsource. The instructor who teaches human qualities like conscientiousness and who motivates his student needs to *be there*.

In formal systems of education, such as colleges and universities, the professor is the center of the instruction and the computer is the add-on. Email the professor with a question. Watch an auxiliary lecture by a Nobel laureate on YouTube. Practice the homework problems on Aplia, a website that can tell you whether you have shifted the supply-and-demand curves the right way. Those features of contemporary education are all very nice, but they also show how far the computer revolution has to go when it comes to education.

Could an education in economics ever be a matter of sitting down with a computer and playing with it? You would manipulate models and verbal problems about the real world. You would solve problems or write out or speak out answers to verbal problems. The computer would correct you and show you better answers, or applaud your acumen. This process would continue for hours and indeed for years. At the end of it all, you would be well equipped to be in a "Freestyle economist human–machine team." The program makes you more productive and it's not that heavy to carry around. If you were sitting in a policy meeting, or doing a consulting job, you would ask your computer program the appropriate economics question, process its answer, and improve upon that answer, submitting your improvement to the computer for feedback and checking. You would have learned how to work with this program very effectively.

I think that one day higher education of many kinds will be substantially a matter of just that. But that isn't the whole story.

Along the way, the budding economist would work with some human instructors. Those individuals would show the student how to best learn from the computer. They would also motivate the student, provide role models, and "bring economics to life." In other words, in the longer run, professors will need to become more like motivational coaches and missionaries. The best professors have understood this for years and have been serving that function from the beginning. What's less well understood is that improvements in AI will make these *the remaining* roles of what we now call "professors." The professor, to survive, will have to become a motivator and coach in essence and not just accidentally or in his or her spare time.

At a good teaching school, a professor is expected to run the class and, sometimes, have a small group of students over to his house for dinner. As the former function becomes less important, due to competition from online content, the latter function will predominate. The computer program cannot host a chatty, informal dinner in the same manner. We could think of the forthcoming educational model as *professor as impresario*. In some important ways, we would be returning to the original model of face-to-face education as practiced in ancient Greek symposia and meetings in the agora.

It will become increasingly apparent how much of current education is driven by human weakness, namely the inability of most students to simply sit down and try to learn something on their own. It's a common claim that you can't replace professors with Nobel-quality YouTube lectures because the professor, and perhaps also the classroom setting, is required to motivate most of the students. Fair enough, but let's take this seriously. The professor is then a motivator first and foremost. Let's hire good motivators. Let's teach our professors how to motivate. Let's judge them on that basis. Let's treat professors more like athletics coaches, personal therapists, and preachers, because that is what they will evolve to be.

The Mormon Church has been fairly successful at getting a large number of people to convert to the Mormon religion. Let's set the theology aside and see if their methods of motivating converts can teach us anything useful about how we could improve education. If you Google "Mormon DVDs" you will see plenty of them, and at affordable prices, but most conversions are driven by Mormons themselves, either in their role as missionaries or as role models in everyday life. The individual Mormon has influence as a living, walking exemplar of what the convert can aspire to.

Of course, educational institutions aren't ready to admit how much they share with churches. These temples of secularism don't want to admit they are about simple tasks such as motivating the slugs or acculturating people into the work habits and sociological expectations of the so-called educated class. As it currently stands, we are losing track of a college education's real comparative advantage. This was an acceptable bargain when the wages of educators and administrators were low, and government budgets had more slack, but it's becoming increasingly expensive.

We like to pretend our instructors teach as well as chess computers, but too often they don't come close to that ideal. They are something far less noble, something that we are afraid to call by its real name, something quite ordinary: They are a mix of exemplars and nags and missionaries, packaged with a marketing model that stresses their nobility and a financial model that pays them pretty well and surrounds them with administrators. It's no wonder that this very human enterprise doesn't always work so well.

Of course, not all students or workers are like chess players who are fine with solitary sedentary activities and well motivated, as no one is forcing them to play. In comparison, college students need more than a little motivating, along with some compassionately delivered but brutally honest assessments of their accomplishments. It is actually quite easy to imagine that we might put the world's very best education online, for free, or for the costs of a few apps, and that most of the world—the undereducated most of all— wouldn't really care very much.

What does the resulting model of education look like? The better-performing students will be treated much as chess prodigies are today. They will be given computer programs to play with, with periodic human contact for guidance, feedback, and upgrades to new and better programs. They will cooperate with each other toward the end of greater mastery of their subject areas. Their conscientiousness, and the understanding that high wages await them in the world, will enforce hard work and discipline.

The lesser-performing students will specialize in receiving motivation. Education, for them, will become more like the Marines, full of discipline and team spirit. Not everyone will adopt the so-called "tiger mother" or Asian parenting style, but its benefits will become more obvious. A lot of softer parents will hire schools and tutors to do this for them. The strict English boarding school style of the nineteenth century will, in some form or another, make a comeback. If your eleven-year-old is not getting with the program, you will consider sending him away to the hardworking, whip-cracking Boot Camp for Future Actuaries. Neo-Victorian social ideals may not triumph, but they will become a much stronger force among lower earners.

We already have the KIPP schools, which stands for Knowledge Is Power Program. KIPP is a nationwide network of open-enrollment schools, mostly for underprivileged youths. Studying in KIPP is not easy. The school insists on longer hours from its students, often running 7:30 to 5:00 Monday through Friday and 8:30 to 1:30 on select Saturdays. There is often mandatory summer school, and in general the classes are much harder than what an underprivileged student in the United States might otherwise face. An overwhelming majority of the students are either Latino or African American. A student is accepted only after a mandatory home visit, after which the student and the parents must sign a contract pledging they will do everything possible to help the KIPP agenda and eventually send that student to college.

The studies to date indicate that KIPP schools have brought very real successes to their students, and that is after adjusting for the fact that KIPP schools are likely attracting more ambitious students in the first place. The point is not that KIPP will work for all students, but rather that it will work for some; remember, *average is over*.

As well as boot camp for students, boot camps for teachers are proliferating. In Boston's Match Teacher Residency program, potential teachers go through a hundred hours of drills with role-played students. Amanda Ripley observed in an *Atlantic* magazine piece:

> The Institute for Simulation and Training runs a virtual classroom at 12 education colleges nationwide— using artificial intelligence, five child avatars, and a behind-the-scenes actor. Some trainees find the simulation so arduous that they decide not to go into teaching after all.

Machine simulations and all kinds of mechanical intelligence aren't making getting to the top easier. But they are making the results better, for both students and teachers.

Especially charismatic teachers will surely have their place—and probably a very well-paid place—in the new world of work. Hong Kong already has glamorous celebrity tutors, called "tutor kings," who help prepare students for the all-important scholastic exams. These instructors are good-looking, photogenic, and their personal lives turn up in the celebrity gossip papers. The female tutor kings dress up stylishly, perfecting their makeup and refusing to wear the same outfit twice. It is rumored that Richard Eng, one of the leading tutor kings, pulls in $1.5 million a year; his face is on billboards, he drives a Lamborghini, and his license plate reads simply "Richard." There has been no comprehensive study of the effectiveness of these tutors but it is believed they do a better job getting the students to pay attention to the lessons. Who could say no to Natalie Portman or LeBron James? Doing well on the Hong Kong exams at age sixteen can mean the difference between getting into a top school or being left behind to stagnate in an ordinary career.

In general, the higher the wage premium for the educated, the more individuals will be willing to endure hardship and severe discipline, or shell out large sums of money, to cross over to the other side of the cultural divide and join the ranks of the highly educated. For these individuals, elevating their psyche to a higher level of ambition will bring big bucks and much higher social status. Parents will increasingly require their children, from a quite young age, to use intelligent machines that test which techniques work best on each individual kid.

It's an open debate how much education can boost innate aptitude or IQ, but the trait of "conscientiousness" does consistently predict educational and job success and also subjective happiness. Yet as access to information increases, conscientiousness will become all the more important. It will be less about whose parents could afford Harvard or who could charm the admissions officer, and more and more about who sits down and actually starts trying to master the material. And so a large part of the educational sector will be directed toward boosting conscientiousness, though not always with success.

When a person is not doing what he or she is supposed to be doing, someone has to deliver that message in just the right way. Show up on time! Don't shop online at your desk! Sell more of our products! Listen more closely to our customers! It is a complicated communication because you are both making the person feel bad about what they have been doing and getting them willing to achieve better results. Expert coaching or motivating will be a competitive growth sector for jobs.

And just as conscientiousness will become a more important quality in labor markets, so will teaching and instilling conscientiousness become more important in the economy as a whole, a theme outlined by Daniel Akst in his brilliant yet neglected 2011 book *We Have Met the Enemy: Self-Control in an Age of Excess*. A lot of new jobs will be coming in the area of motivation. These jobs will require some very serious skills, but again they won't primarily be skills of a high tech nature or skills that are taught very well by our current colleges and universities. And again, these high expertise coaching jobs won't be shipped overseas.

High-skilled performers, including business executives, will have some kind of coach. There will be too much value at stake to let high performers operate without a steady stream of external advice, even if that advice has to be applied rather subtly. Top doctors will have a coach, just as today's top tennis players (and some of the mediocre ones) all have coaches. Today the coach of a CEO is very often the spouse, the personal assistant, or even a subordinate, or sometimes a member of the board of directors. Coaching is already remarkably important in our economy, and the high productivity of top earners will cause it to become essential.

At various career steps, individuals who work with genius machines will need to retrain and learn new systems. Some will opt for self-education, supplemented by programs and some human guidance, much like the chess prodigies. Those who are less self-motivated will subject themselves to extreme forms of discipline for short periods of time, to learn a new set of skills. And others will retreat into the world of what I have called threshold earners, just trying to get by.

Today's leading chess Freestyle players have no formal education in that craft, but they often come from technical backgrounds and are strong on self-motivation, in addition to being intimidatingly bright. Before his current work on building up an openings book for his Freestyle chess team, Nelson Hernandez (now in his midfifties) worked as an army paratrooper, a stockbroker, for a hedge fund, and as a financial analyst, the job he now holds. He describes being good at "dull, repetitive tasks" and "really wanting to win" as making him well suited for his Freestyle avocation. Those are qualities that will serve well the workers of the future.

Larry Kaufman, who developed the evaluation function for the Rybka program, and who is the mastermind of the Komodo program, graduated from MIT with an undergraduate degree in economics in 1968. He went to work on Wall Street as a broker and soon started developing his own form of options-pricing theory, working independently of Fischer Black and Myron Scholes; Scholes later won a Nobel Prize for that contribution. Kaufman's theory was based on ideas of Brownian motion and the logistic function, the latter of which he took from formulas for calculating chess ratings. In the 1970s he made money by applying his options-pricing work through a trading firm and stopped when the profits went away, and he has since dedicated his life to chess and computer chess, including his work on Rybka and Komodo. He lives in a fine house in one of the nicest parts of suburban Maryland, with his beautiful wife and young daughter. Again, we see the mix of a moderate level of elite education combined with extreme self-education over many years. In his midsixties, Kaufman is still making pioneering contributions to the theory and practice of intelligent machines. He and Nelson Hernandez are good examples of the kind of skills that will win out in the future. Above all else, they are masters of reeducation.

11

The End of Average Science

Many of us have striven to work at something that is not only well paid but is meaningful and important. We want to contribute something substantial. Some of us wanted to be teachers, some medical doctors, some particle physicists. In the vast array of career choices it is easy to overlook the fact that modern professions all depend on scientific discoveries to one degree or another. So what about science? Is average over for science?

Science is a general framework for making predictions, controlling our environment, and understanding our world. I believe, however, that the practice and understanding of science is in for some major changes. And the reason is probably not a surprise: mechanical intelligence. We eventually will cease to understand significant parts of the science underlying our jobs and lives—many of us have already. Counterintuitiveness and an inability to understand are already common in quantum mechanics. That trend toward incomprehension will continue.

We are standing at an unusual point in the history of science: Most important scientific results can still be understood by reasonably educated human beings and certainly by well-trained, intelligent researchers. A lot of the popular books on science are pretty enlightening. They probably don't give you the ability to adjudicate potential advances in the field, or appreciate all of the nuances. Still, many Americans who aren't scientists can follow some of the basic results of, say, evolutionary biology or Einstein's theory of general relativity.

We should not, however, take that state of knowledge as fixed. I'm not talking about a decline in literacy here—science itself is, in many areas, moving beyond the frontiers of ready intelligibility. For at least three reasons, a lot of science will become harder to understand:

1. In some (not all) scientific areas, problems are becoming more complex and unsusceptible to simple, intuitive, big breakthroughs.
2. The individual scientific contribution is becoming more specialized, a trend that has been running for centuries and is unlikely to stop.
3. One day soon, intelligent machines will become formidable researchers in their own right.

The overall picture is a daunting one for the ability of the individual human mind to comprehend the science of how our world works.

Specialization

As science progresses, each new marginal discovery is more the result of specialization and less the result of general breakthroughs, compared to earlier times. There probably won't be another Isaac Newton, Adam Smith, or Euclid, because the most fundamental contributions in those fields have already been made. New fundamental contributions are hardly over, but they will come in dribs and drabs and they are more likely to come from research teams than from lone geniuses in major, unexpected bursts. There is nothing wrong with that, and in fact it reflects some positive features of science, namely that communications are rapid and intense, that there are many very talented people working on the major open problems, and that a lot of basic progress is already behind us. Science is more of a cooperative endeavor than in earlier times, and that means the individual research contribution is smaller, even during periods of great progress.

We're already at the point where there is not always common agreement as to what it means to "prove" a mathematical theorem. An important theorem could fill dozens or hundreds of pages and rely on hundreds of prior results from different parts of mathematics. The theorem as developed relies on a division of labor, but quite literally no single mind knows if the theorem is true and instead a group of mathematicians goes over the theorem, divvying out the parts to the appropriate specialists. There is a collective judgment of proof, or not, and only then does the innovator discover if she has come up with anything important.

In 2010, researcher Vinay Deolalikar from HP Labs claimed to have a proof of the famed mathematical proposition that P is not equal to NP, one of the famed Millennium Prize Problems in mathematics and worth a $1 million prize. Even he didn't know at first whether he had succeeded. He published the supposed proof in the form of a hundred-page paper on the internet, open to the scientific community. Even a year after the publication, it was not totally clear if it was indeed a proof. Many mathematicians were skeptical and Deolalikar admitted that his original proof had problems, but nonetheless he redid the work and claimed success in the revisions. Specialists tackled each different area of the proof, until eventually the skepticism grew and seemed to take hold. As I write, the matter is still unsettled, though the mathematical community is leaning in a negative direction against the proofiness of the supposed proof.

Grigory Perelman did better. He was awarded a Millennium Prize of $1 million in March 2010 for his proof of the Poincaré conjecture, which states that "every simply connected, closed, 3-manifold is homeomorphic to the 3-sphere." If you think that summary is hard to understand, just try to figure out the proof. The original submitted proof had arrived many years earlier, in a series of papers from 2002 and 2003, but again, at first no one was sure if he had succeeded. (By the way, Perelman eventually turned down the prize, saying he didn't want money or fame.)

When it comes to complicated proofs, there is no single mind that understands the truth of the theorem or what the theorem really means, if indeed that concept is well defined in the first place.

Specialization is also reshaping applied science and invention. Formerly, a researcher or potential inventor could learn the entirety of a scientific or applied area in a few years' time, master it, and produce an innovation rather quickly, often working alone or in a very small group. The major inventions behind the Industrial Revolution, for instance, were often driven by amateurs. That's become a lot harder because there is so much knowledge to master in the mature fields. It can take ten years of study or more to get to the frontier of a lot of areas, and by the time you get there, and figure out something new, your contribution is a marginal one or maybe a little out-of-date. The frontier moved on while you were trying to master it. Even if you succeed, you'll understand why your tweak is better than the way things used to be done, but your understanding of the new device as a whole may be rudimentary or even incorrect, because you relied so much upon the underlying knowledge of others.

There are exceptions to this rule in fields that are not yet mature. As I've already cited, in social networks software Mark Zuckerberg was a pioneer while still an undergraduate at Harvard. It didn't take him long to get to the frontier of social networking, and he quickly redefined that frontier in a fundamental way. He had plenty of help with Facebook, and he built upon earlier social networks such as Friendster and Myspace, but to a large extent Facebook reflected his personal vision and expertise. At least in the early years of the product, Zuckerberg had a comprehensive bird's-eye view of what Facebook was all about. Online social networks were new at the time and so fundamental contributions were easier to achieve.

The ability to "go it alone" is conducive to rapid innovation and innovation by amateurs. Lone individuals and small groups can make major contributions, and that limits the stultifying effects of bureaucracy and regulation. Zuckerberg needed some initial aid and financing, but not many people were in a position to tell him, "No, you can't do this." Since his initial breakthrough, Zuckerberg has shown the ability to take a long-term perspective on the future of the product and change course when needed. Such are the benefits of small-group innovations. Now "writing good programs for social networking" is becoming a more mature endeavor and designers are specializing more and more in the tweaks. Social networks, considered as a whole, are becoming less comprehensible to the individual expert mind, Zuckerberg included.

Most areas of science and applied invention do not have much room for Mark Zuckerbergs, because the fundamental break-throughs have been in place for a longer period of time. It's hard to sit around in your garage, or campus dorm, and figure out a new and better way to design an automobile. Nonetheless, there is a kind of race going on. As scientific progress makes lone contributions harder to sustain, machine intelligence may be creating new pos-sibilities for the amateur and for the nonconformist. A genius ma-chine can play the role of an investigator on a scientific team. Let's say you can't test your idea for a better tire design on an actual rain-slicked road with an expert driver and a padded vehicle. Maybe a computer simulation can help you make progress even without the resources of General Motors at your disposal.

At any point in time, there will be some new breakthrough ar-eas, where innovators are fundamental and redefine entire sectors of knowledge rather quickly and perhaps hold much of that knowl-edge in their heads. Still, as the accumulated total of human knowl-edge increases, those breakthrough sectors become just a small part of our scientific understanding of the world. Science tends to look more like bureaucracy, and in standard bureaucracies no single mind has much of a grasp of the whole. In my own field, economics, coauthored pieces are already becoming much more common, and with a greater number of authors, as they are in many other fields of science as well.

Eventually, dare I say it, science will also look more like religion and magic because of its growing inscrutability. The working parts will be hidden, much as an iPhone functions without showing you its principles of operation. You will be able to see the bureaucracy if you look at the scientific community, and every day you will be exposed to the magic as a customer or on your job.

Impossible Problems

There is a further reason why many of the sciences will become harder to understand.

Thirty or forty years ago, it might have been rational to hope that such tough areas as cosmology, fundamental physics, genetics, and even macroeconomics would fall into place with some simple, easily rationalized, and deeply compelling general approaches. Einstein's general theory of relativity, for instance, though counterintuitive at first glance, offers a relatively simple structure. Once you get it, you get it and you can think in those terms. You can debate the paradoxes of time travel and understand some of the gimmicks they put in science fiction movies, or recognize when those gimmicks botch the basic science. Some of us can even write it down in terms of equations.

In more recent times, in many particular areas, the hopes for comparably simple major breakthroughs have been dashed on the rocks. There have been plenty of scientific advances, but the world seems to be a messier place conceptually than before. Genetic explanations for human behavior continue to grow, but the connection between genes and outcomes is growing messier and more complicated all the time. Even the height of a person—a clearly heritable characteristic—seems to involve dozens of distinct genes, with more being found all the time. We're not going to find a "gay gene" or an "autism gene," even though genes play major roles in both homosexuality and autism.

Or consider the recent detection of the Higgs boson. On one hand it held out the promise of completing previous particle theories and tying up their loose ends with some empirical verification. On the other hand, researchers are already wondering what deeper truths of a "Grand Unified Theory" might lie underneath our current understanding. The candidate approaches here are hardly simple or intuitive, and it's far from obvious that they will converge to general intelligibility over time.

We simply may have reached the point in some key scientific areas where we are working with levels of explanation that our human brains—even those of Nobel laureates—cannot handle. The top scientists might end up being people not who "know," but rather who hold shadowy outlines of the truth in their heads. Today's frontier questions about cosmology, epigenetics, or macroeconomics are all more complicated and more advanced than the questions being asked forty or even twenty years ago. There is no guarantee that future advances will move us back to simpler conceptual worlds, and if anything the likelihood seems to point in the opposite direction. If we increasingly rely on the genius of machines to crunch a lot of data, in lieu of a simple and easily intelligible overarching framework, will that really be so bad?

Due to advanced mathematics, we are already developing theories that very few people understand, if indeed any people understand them at all. String theory, made somewhat familiar by Brian Greene in his hugely successful popular-science book *The Elegant Universe,* is far from intuitive. Perhaps no one really knows what it means to postulate ten or more dimensions. We can handle those dimensions with advanced mathematics, but it is an example of how a scientific theory can evolve to the point of unintelligibility. Just try this description. It is taken from the first page of the Wikipedia entry, which of course is trying to be as accessible and as intelligible as possible:

> String theory posits that the electrons and quarks within an atom are not 0-dimensional objects, but made up of 1-dimensional strings. These strings can oscillate, giving the observed particles their flavor, charge, mass, and spin. Among the modes of oscillation of the string is a massless, spin-two state—a graviton. The existence of this graviton state and the fact that the equations describing string theory include Einstein's equations for general relativity mean that string theory is a quantum theory of gravity. Since string theory is widely believed to be mathematically consistent, many hope that it fully describes our universe, making it a theory of everything. String theory is known to contain configurations that describe all the observed fundamental forces and matter but with a zero cosmological constant and some new fields. Other configurations have different values of the cosmological constant, and are metastable but long-lived. This leads many to believe that there is at least one metastable solution that is quantitatively identical with the standard model, with a small cosmological constant, containing dark matter and a plausible mechanism for cosmic inflation. It is not yet known whether string

theory has such a solution, nor how much freedom the theory allows to choose the details.

That was the easy part. Now:

String theories also include objects other than strings, called branes. The word *brane*, derived from "membrane," refers to a variety of interrelated objects, such as D-branes, black p-branes, and Neveu–Schwarz 5-branes. These are extended objects that are charged sources for differential form generalizations of the vector potential electromagnetic field. These objects are related to one another by a variety of dualities. Black hole-like black p-branes are identified with D-branes, which are endpoints for strings, and this identification is called Gauge-gravity duality. Research on this equivalence has led to new insights on quantum chromodynamics, the fundamental theory of the strong nuclear force. The strings make closed loops unless they encounter D-branes, where they can open up into 1-dimensional lines. The endpoints of the string cannot break off the D-brane, but they can slide around on it.

This forbidding summary doesn't have to imply anything negative about the theory; on the contrary, it reflects how advanced tools can provide some understanding to theories beyond easy comprehensibility by most, nearly all, human beings. Without advanced mathematics, string theory could not have been formulated in the first place.

These days, in science, in a variety of fields, it is common for researchers to write down equations that very few people understand. It's not much of a leap to think we will arrive at the point—or maybe we already have—where *no one* understands the equations being written down. Understanding is of course a matter of degree, and we can imagine the leading scientists in an area understanding a successively smaller part of the results they are producing. This is possible because the production of the results comes from a team. The other parts of the understanding will be held in the minds of other researchers or generated by genius machines. It's similar to how no one person on the assembly line understands very much about how an automobile works, nor do they need to. Just as Adam Smith and Friedrich Hayek and Michael Polanyi stressed that a market economy evolves to the point where it is very difficult to understand the overall interrelationships of production, so can the same be said for many branches of science.

Overall, the difficulties of grasping the big picture are reflected in the age structure of scientific achievement. Arguably, older scientists are less bold, less innovative, and conceptually more set in their ways. Einstein once said, "A person who has not made his great contribution to science before the age of thirty will never do so." That's no longer true (if it ever was), but still Einstein had a point. Innovators are often freshest and most revolutionary in their younger years, and with some age we acquire wisdom but we lose some of the sharp conceptual edge and the willingness to overturn established ways. The innovators we end up with tend to be less revolutionary, again with the notable exception of the internet and internet-related innovations, where young people can still move to the frontier very quickly. The older and usually middle-aged researchers are nonetheless responsible for more breakthroughs, because only they have enough knowledge to grasp a reasonable whole.

Researchers Bruce Weinberg and Benjamin Jones analyzed the 525 Nobel Prizes awarded in physics, chemistry, and medicine from 1900 to 2008. The trend in all fields is that the pathbreaking researchers are older, when they do the prizeworthy work, as the decades pass.

In 1905, the typical Nobel-winning physicist made his or her breakthrough at age 37 but by 1985 the average age was 50. The average age for a Nobel-worthy chemistry breakthrough, over that same time period, rose from 36 to 46 and for medical scientists it rose from 38 to 45. Prior to 1905, 20 percent of these Nobel Prizes were won for discoveries before the age of 30, but by 2000 hardly any Nobel Prizes resulted from the work of such prodigies. Whether we like it or not, breakthrough science has passed into the hands of the middle-aged.

These developments may prove problematic for areas such as mathematics, which have relied relatively heavily on prodigies. Better access means it is easier to have the knowledge of a math prodigy today, yet it is harder to be a groundbreaking math prodigy. Perhaps by the time you get to the frontier of knowledge at thirty years old you have lost some of your edge.

We need to come to terms with these trends, but we probably cannot reverse them and perhaps we should not even try. As I've mentioned, the status quo, and its likely future directions, has many factors in its favor, including greater ease of scientific communication and collaboration, greater access to scientific materials, greater computing power, intelligent-machine proliferation, and the greater number of people who have a chance to do science, including from China and India. The benefits from those developments will, in absolute terms, likely offset the problems of specialization. In any case, there is no way to return to the days of Euclid, where an entire field can be revolutionized or created by a single book or set of lecture notes. This is fundamentally a story of progress, but it will be an unusual kind of progress, operating in large part above and beyond the normal channels of human understanding.

One problem with this kind of progress is that it is hard to regulate. I don't just mean regulate by the government, but rather regulate in the broader sense. It will be increasingly hard for scientist administrators, philanthropists, and also government bureaucrats to get a handle on what is going on in a lot of scientific areas. The inscrutability of science will place an increasing burden on trust, whether it be trust in particular institutions, scientists, or reward structures such as the Nobel Prizes. How about trust in Google? The overall wisdom will reside in the system of science rather than the individual mind, but that will be problematic when particular individual minds need to make decisions about how to allocate resources within the system. Future science will be much more of a spontaneous order and much less of a planned or easily visualized community with principles that can be easily articulated or explained.

Machine Science

Most current scientific research looks like "human directing computer to aid human doing research," but we will move closer to "human feeding computer to do its own research" and "human interpreting the research of the computer." The computer will become more central to the actual work, even to the design of the research program, and the human will become the handmaiden rather than the driver of progress.

An intelligent machine might come up with a new theory of cosmology, and perhaps no human will be able to understand or articulate that theory. Maybe it will refer to non-visualizable dimensions of space or nonintuitive understandings of time. The machine will tell us that the theory makes good predictions, and if nothing else we will be able to use one genius machine to check the predictions of the theory from the other genius machine. Still, we, as humans, won't have a good grasp on what the theory means and even the best scientists will grasp only part of what the genius machine has done. It would be like trying to explain the Periodic Table of Elements to a five-year-old. Maybe it can be done, but it will hardly serve as an easily grasped, intuitive understanding of what is going on.

The incentives for producing better science will encourage this broader unintelligibility. Machine intelligence is less valuable for performing the tasks and calculations that humans can already understand fairly well. They may do those tasks more rapidly, but a lot of the potential gains come from using the machines to do things that humans cannot much handle or comprehend at all. That's the division of labor and complementarity, both of which can push scientific results away from general intelligibility once those genius machines enter the game.

The resulting kind of unintelligibility will depend on the scientific area. Some disciplines, such as cosmology, attempt (among other things) to construct grand, all-encompassing theories. There is the possibility that no human will understand the best grand theory on tap, because that theory is too complex or too advanced or because the categories of the theory are sufficiently far removed from our everyday experience.

Much of normal science doesn't fit this pattern, such as when researchers gather and refine data about the digestive system of a particular kind of starfish, or when they study the lava flow of a volcano, or many other examples. In those cases, there probably isn't any new grand theory in the offing, genius machines or not. Instead there will be more data gathering, more hypothesis testing, and a slow refinement and improvement of existing knowledge. There will be a growth in the division of knowledge. It will be much harder to have a comprehensive knowledge of the current science of the digestive system of the starfish, although through the internet it will be easier to call up any particular piece of knowledge in an area. There will be lots of "micro-intelligibility" on tap, although less big-picture thinking in ways that are accessible to nonspecialists.

The remaining human knowledge of science will be very practical, very prediction-oriented, and well geared for improving our lives. Of course those are all positive developments. Still, as a general worldview, science will not always be very inspiring or illuminating. The general educated public will to some extent be shut out from a scientific understanding of the world, and we will run the risk that they might detach from a long-term loyalty to scientific reasoning.

On the brighter side, the better educated segment of the lay populace is already moving back to a closer involvement with *doing* science, even if the involved individual doesn't have a great grasp of the overall theory. Think of amateurs using their telescopes and computers to search the sky for new discoveries, whether or not they have a good grasp of supernovas and black holes. Think of the importance of amateur bird-watchers for ornithological data and studies. Think of lay users who contribute parts of their home computing power to scientific projects—yes, this is already happening.

Individuals are also contributing personal data about their health, diet, and the behavior of their pets to enable scientific and medical studies. The assembling and manipulating of the vast swathes of data that result from these projects is called "citizen science" and it is a growing trend. It does not depend on the individuals fully understanding the hypotheses being tested.

When it comes to the relationship of the intelligent citizen with science, we will increasingly become doers and participants rather than comprehending observers.

As it stands now, we don't have much in the way of grand theories that are beyond human intellectual capabilities. Since only humans come up with theories, by construction the theories are within the intellectual grasp of at least some humans, albeit humans who are usually smarter and better educated than average. Once genius machines start coming up with new theories, that constraint will be removed and someday intelligibility will seem like a legacy from a very distant past. We should not assume that we currently know which scientific areas will offer the mundane data gathering and which will offer the exciting grand theories from the genius machines. The digestive system of a starfish might be exactly the kind of area where the machines can see regularities that we do not, and come up with some suitably complex—but to us unintelligible—theories.

We will see, more and more, the relatively mundane data-gathering sides of science. The bureaucracy and data gathering of science will be visible, as will be the magic of the devices we use. But that middle layer of knowledge—science as a general means for educated laypersons to understand the world through theories—will peak sometime in the twenty-first century.

Whither Economics?

Academics tend to be provincial with respect to their fields and specialties, so allow me to focus for a moment on my particular science. The so-called dismal one.

In the last ten years there has been a big shift in emphasis and it has come largely from web companies, not from academic researchers. When web companies are figuring out their business models, and trying to market to their customers, they tend to use a lot of raw, relatively unfiltered data. Quite simply, they do this because they can. Facebook, Google, Amazon, and other companies have a phenomenal amount of high-quality information at their disposal, more than most academic economists are used to having. And when they process this data, they go a relatively atheoretical route. They "crunch" the data, and we now have "Big Data," as we've come to call it, as the next business revolution, which refers to the use of statistics on the data generated by electronic communications.

These companies, in their approach to this data, are fairly suspicious of structural theoretical models. They think about coding the data properly and organizing it in useful ways, but they're not trying to start with "the Jonesian model of why people use Google," or "the Brownian model of which books people buy on Amazon." They go straight to the numbers and try to find power where they can.

Economics as a research area, in recent times, has been following the same path as these web companies: lots of data and relatively weak theoretical structure. Powerful data crunching, and careful data gathering, is pushing out theoretical intuition. We still haven't dispensed with models, because there are a few models we believe in pretty strongly, such as that when price goes up, people usually buy less of that good or service, all other things being held equal. But those are old theories and the real action and value-add comes from the data and its handling, including data from field experiments, laboratory experiments, and from randomized control trials. The underlying models just aren't getting that much better, and when the underlying models are more complicated, they very often are not more persuasive to the typical research economist.

I would sum up the blend as follows: (a) much better data, (b) higher standards for empirical tests, and (c) lots of growth in complex theory but not matched by a corresponding growth in impact. Mathematical economics, computational economics, complexity economics, and game theory continue to grow, as we would expect of a diverse and specialized discipline, but they are if anything losing relative ground in terms of influence. Economics is becoming less like Einstein or Euclid, and more like studying the digestive system of a starfish.

If there are any economists who are having a big impact these days, it is Esther Duflo and Abhijit Banerjee, and their colleagues at the Poverty Action Lab at MIT. I once visited one of their research projects in Hyderabad, India. It involved tens of thousands of subjects, some of whom had access to microcredit and some of whom did not. The two groups were drawn from roughly comparable neighborhoods and the goal was to compare how big a benefit microcredit really was, or not. An army of dozens of assistants helped gather data from the borrowers, both before and after they started (or did not start) the microcredit program. That included data about income, new jobs or businesses, failure to repay loans, and many other features of their daily economic lives. The basic question was a pretty simple one: whether the group with access to the microcredit did better. It turned out they were more likely to have started their own businesses and thus a classic paper was born. Most people see this as the most important study of microcredit, in addition to another large-scale randomized control trial from Dean Karlan at Yale University. It's a long way from grabbing a publicly available database from a government agency, without much worrying about the quality or meaning of the numbers, and running some regressions. Setting up the entire field experiment is also a uniquely human contribution and it does not approximate any task that is replicable with smart machines.

Outside of economics, a computer program will look at a lot of numbers, search for patterns in a more complex way than current empirical researchers can do, and report back the results. You can imagine culling a lot of profiles from social networks and seeing how much taste in music is explained by gender, age, and where the person lives. These programs will confirm some connections we already believe in, see some connections that we currently do not grasp, and perhaps generate some hypotheses that we do not suspect. Economics is not yet there, but perhaps in the next fifty years such endeavors will supplant the economist's reliance on theoretical models. The power and quality of data will likely grow more rapidly than the power and quality of our best models.

Current model-building in the social sciences is analogous to "grandmaster intuition pre–Deep Blue." Making models has been a very useful approach and indeed it still is useful because the Deep Blue of the social sciences has yet to arrive.

In economics, the early uses of machine intelligence will reinforce our understanding of some basic regularities behind economic phenomena. We will come away with a renewed and more

accurate sense of what precedes a financial crisis, what predicts excess stock returns, or which cultural factors precede economic development. We will feel better about what we think we already
know, along with some revisions to knowledge at the margins. In
the much longer run, as data quality improves and the number of
data points multiplies, machine intelligence might tell us that a particular mix of regulations and monetary policy will lead to a financial crisis (with some specified degree of certainty, of course) and
we might not see why. We will look at the machine's reasoning but
it will be too data-rich and the models will be too complex for us to
grasp readily. We will know how to feed the machines with data,
and how to test them against each other, and we will know how to
use their results. But at some point we will cease to understand all
of the component parts of the science and we will cease to understand how the predictions are put together. Only the machine will,
in its own way, be able to encompass the entirety of the theory and
its tests.

The machines will eventually encroach upon all or most of the
functions of the economist. The social scientist of the future will no
longer be an independent agent who formulates theories, tests them
against data, and writes up the result for publication. The social scientist of the future will team up with computing power to an increasing degree, specializing in complementing the progress driven
by the programs. Some notion of publication may still exist, but the
important outlet for research will be in standardized, machine-
digestible form. Rather than "reading articles," we will consult the
programs to spit out the results of their meta-studies, summarizing
the research work to date, much as Rybka spits out an evaluation of
a chess position. What used to be an individual journal article will
become an input into the programs. The "expert" might be someone

trained in making sense of the machine's output, or turning the data into machine-readable form, rather than someone who does the actual work generating the estimates.

That's the single biggest change in economic science we can expect over the next fifty years. When it comes to "the new paradigm," a lot of people are expecting the next Marx, Keynes, or Hayek. The changes to come will be more radical than that and they will challenge the very relationship that the scientist has to his or her craft of study. The real change will be the subordination of the individual scientist.

I see early indications of these trends in economics, one of the branches of the social sciences most influenced by computers. Newly minted PhD candidates are extremely proficient with data, but a lot of them don't have much microeconomic intuition. You could ask them some simple microeconomic questions, of the kind the University of Chicago used to pose at the undergraduate level, and not get much of an answer. If you ask job market candidates with newly minted PhDs, "Under what conditions will allowing brands to purchase shelf space in supermarkets, as opposed to banning the practice, benefit customers?" you will end up with a fair number of blank stares. That's a question of pure microeconomic logic, and it's pretty basic in structure (which is not the same thing as being easy), but those skills aren't taught very much anymore. These same people who fail that question of microeconomic intuition may be quite proficient at computer programming or massaging data into usable form. Overall, the profession is producing more first-rate empiricists than before, yet theory hasn't progressed much in twenty years or more. Theory is increasingly ignored.

If anything, the seminal work in fields such as development economics and labor economics—two data-driven fields, and two fields that are growing in importance—is in using simpler theories than was the case twenty or thirty years ago. It's the simpler theories that allow us to harness the ability of computers to analyze data, or that allow us to set up ready-to-use field experiments. In this regard, the path of economics is very different from that of theoretical physics or cosmology. The theories in economics that have actual resonance are becoming simpler, and the more complicated theories, while they are still around, are losing general influence.

Today most of the debate on the cutting edge in macroeconomics would not call itself "Keynesian" or "monetarist" or any other label relating to a school of thought. The data are considered the ruling principle, and it is considered suspect to have too strong a loyalty to any particular model about the underlying structure of the economy.

If I see an important economics paper, circa 2013, odds are it was based on a clever way to find or generate a new data set, not a new theoretical idea. Data gathering, of course, complements data crunching, and it is something that machine intelligence is not close to managing. The computer cannot speak to people in a Rwandan village or understand what questions to ask, much less record the answers and turn them into a usable form.

One way to see the melting away of the theoretical apparatus is to look at a few of the leading careers in recent times. Steven Levitt writes papers on baby names, sports, and whether teachers cheat, classic topics in sociology education research and other areas. Nobel laureate Gary Becker has spent decades studying the family and household behavior, even when traditional exchange and dollars-and-cents reasoning does not apply. Daniel Kahneman, who is a psychologist, has won a Nobel Prize in economics, and more economics is coming out of law schools. Paul Krugman has very explicitly declared his primary allegiance to the simpler models. A lot of the pioneering work in political science has been done by economists, under the "public choice" or "political economy" moniker. We're not far away from having a single de facto, more or less unified, empirical social science. In that social science, researchers invest a lot in learning empirical techniques and then invest some marginal energies in the simpler theories that surround their chosen field of study. Finally, they spend their research time looking for new data sets, or looking to create that data, whether by detective work or by lab and field experiments.

The economists who favor more intuitive approaches will take a different tack to survive in the profession. They will specialize less in producing original research, but they will instead become clearinghouses for and evaluators of the work of others. They will translate that work, not just for a broader public but for members of their own economics profession. In essence, these individuals will sit at their computers, much as Anson Williams the Freestyle chess player does, and digest inputs from diverse sources. They will hone their skills of seeking out, absorbing, and evaluating this information. When it comes to judging the truth of a given economic proposition or "move," to borrow an analogy from chess, they will have remarkably high skills, higher than those of many Nobel laureates, despite their absence of first-tier research accomplishments. They will be translators of the truths coming out of our networks of machines.

These Freestyle researchers will be pioneering a fundamentally new way of "doing" economics and a fundamentally new sense of what it means to be an economist and indeed a scientist. They will earn good money and a degree of public fame, and their numbers will multiply, even as their daily routines become increasingly estranged from the practices of normal everyday science in their fields.

At least for a while, they will be the only people left who will have a clear notion of what is going on.

12

A New Social Contract?

What will become of America as a whole in twenty to forty years? What will our politics look like in the new world of work?

One is reminded of the old saw attributed to Yogi Berra: "Prediction is difficult, especially about the future." Still, it can help us make sense of the present to lay down some possible trends about where things are headed. A lot of people will have serious objections to some of these trends. So be it. Let's first understand where the trends might be coming from.

The forces outlined in this book, especially for labor markets, will force a rewriting of the social contract, even if it is not explicitly recognized as such. We will move from a society based on the pretense that everyone is given an okay standard of living to a society in which people are expected to fend for themselves much more than they do now. I imagine a world where, say, 10 to 15 percent of the citizenry is extremely wealthy and has fantastically comfortable and stimulating lives, the equivalent of current-day millionaires, albeit with better health care.

Much of the rest of the country will have stagnant or maybe even falling wages in dollar terms, but a lot more opportunities for cheap fun and also cheap education. Many of these people will live quite well, and those will be the people who have the discipline to benefit from all the free or near-free services modern technology has made available. Others will fall by the wayside.

The slogan "We are the 85 percent!" probably won't sound as compelling as the Occupy Wall Street version. It will become increasingly common to invoke "meritocracy" as a response to income inequality, and whether you call it an explanation, a justification, or an excuse is up to you. Since the self-motivated will find it easier to succeed than ever before, a new tier of people from poor or underprivileged backgrounds will claw their way to the top. The Horatio Alger story will be resurrected, but only for those segments of the population with the appropriate skills and values, namely self-motivation and the ability to complement the new technologies. It's in India and China that the rise of a new middle and upper class is reflecting this trend most clearly.

This framing of income inequality in meritocratic terms will prove self-reinforcing. Worthy individuals will in fact rise from poverty on a regular basis, and that will make it easier to ignore those who are left behind. The wealthy class will be increasingly self-motivated, will be larger over time, and—precisely because we are selecting ever more for self-motivation—will have increasing influence. It is their values that will shape public discourse, and that will mean more stress on ideas of personal ambition and self-motivation. The measure of self-motivation in a young person will become the best way to predict upward mobility.

We'll also see a lot more of some of the hypocrisies common today. For instance, it's pretty common to hear tenured economics professors at establishment schools espouse the relevance of liberal democratic policies, such as the social safety net. These same individuals, if asked to explain their choice of academic hires, or their choice of which students to push in the job market, often respond in rather harshly meritocratic terms. If a graduating PhD student does not have his job market paper ready by his fifth year of study, it's because "that student didn't have a strong enough work ethic," or something like that. That same professor will be very shy to apply the same kind of rhetoric to discourse about the safety net, for fear of sounding like a non-liberal critic such as, say, Charles Murray. When it comes to a lot of values issues—and what people really believe in their daily lives—the gap between conservatives and liberals isn't nearly as large as it might first seem.

What does that mix of values mean for actual social choices? We'll pay for as much of a welfare state as we can afford to, and then no more.

We've already seen how some pieces of this portrait might fall into place, such as labor market polarization and cheaper education. Let's now think through the role of government. I'll start with public sector budgets, which represent the core of what a government does—that is, raise and spend money.

The Fiscal Crunch

The opening question is simple: If wage growth slows for a lot of Americans, but not all, how exactly will we make ends meet for the American government?

Ever since the debt ceiling debate and the fiscal cliff crises, or earlier, everyone is talking about pressure on budgets and debating higher taxes versus lower government spending. Most of us know that both adjustments will be required, but in any case it will be difficult for our government to make ends meet.

Healthcare costs have been rising about 5 percent a year, or more, on a compounded basis, although with some slowdown coming out of the recent recession. For all the talk that someday this cost inflation must end, it continues apace and so healthcare costs threaten to swallow up the economy. If a sum goes up by 5 percent a year, the resulting magnitude doubles about every fourteen years. It's easy enough to see why it's hard to halt this cost inflation. The American population is aging, retiring earlier, and living longer. Even a tough plan to control healthcare costs is swimming against a very strong stream. The problem could become even larger if people live longer than expected under current estimates or if financial crises, wars, environmental problems, or other major disasters keep down the rate of economic growth.

There is another potential danger. The United States, circa 2012, has been enjoying especially low interest rates for government borrowing, and a spike in interest rates could make the fiscal problem much worse. There is plenty of downside risk in our current fiscal situation, even if it is likely some number of years out. Keep in mind that, as I write, the US government is borrowing about forty cents out of every dollar it spends, an unsustainable state of affairs.

Sooner or later, something has to give.

Economists and policy wonks sit around frequently wondering how the circle will be squared. A lot of them just don't see how the spending cuts and tax increases can be big enough to solve the problem. It would seem to imply either gutting the Medicare program and Social Security or raising taxes to extortionate levels, neither of which will find much political support, even if you thought such moves were good ideas. Yet neither is debt default a real option, as that would create the world's largest financial crisis ever and another Great Depression. Our money markets and our banks would collapse, unemployment would skyrocket, and at the end of that process we would still have to make the tough spending cuts.

It's a common view that "the top 1 percent" can or will fund these forthcoming expenditures by paying higher taxes. I don't think that is likely, for a few reasons, even though I do think the wealthy will end up paying somewhat higher taxes. But why can't they pick up the whole tab? First, with a bit of time, the wealthy are slated to account for a larger share of the economy than just the top 1 percent. The wealthy will grow in numbers, and that also means the wealthy will grow in influence. Imagine that today's millionaires comprised 10 percent of the citizenry; that would make for an extraordinarily influential and politically potent group, much more so than the wealthy today. Can you imagine that group funding the entire future by raising taxes on itself? I don't see it.

Second, it isn't always easy to extract more money from the very wealthy. They have greater access to tax shelters, write-offs, tax deductions, untaxed workplace perks, and overseas accounts, all backed by the best lawyers and accountants money can buy—I suppose machine lawyers and accountants at some point. You might think that all those "outs" will be taken away but they never have been, and I don't see that happening anytime soon. Of course, some earners, usually below the very top, don't have access to those outs, but raising their taxes will be limited for a different reason. For instance, if you are a high-earning lawyer in New York City, and paying state, local, and sales taxes on top of your federal burden, you are already facing marginal tax rates of over 50 percent. There's not that much room to raise rates on a lot of people without tax revenues from those people falling or remaining relatively flat.

The most fundamental point is what economists call "tax incidence." That means you can levy a greater tax on a top earner, but there is no guarantee they will have to bear the actual burden. Let's say we levied a higher tax on J. K. Rowling, the author of the Harry Potter books. She could in turn demand better terms from her publisher and so the price of the books might end up being higher. How taxes get translated into final results on wages, prices, and returns to capital is a complicated topic that economists have studied at great length. We still don't know all the answers but we do know a lot of taxes end up getting passed on to other parties. We also expect that in the future the talented high-earning workers will be in strong demand no matter what, and employers will respond to taxes by raising their pay. What's the point of taxing the top earners if a lot of the burden is passed back on to the ordinary citizen?

So, taxes will go up somewhat, most of all on the wealthy and on capital owners, and our tax code will trim back on some of the more outrageous deductions in the current system. Yet raising taxes is not enough to eliminate our future fiscal problems.

What about spending cuts? The answer doesn't lie there either, though I do expect some expenditures to be cut and indeed they should be cut. But look at the budget: What are the three biggest problems today? It's Social Security, Medicare, and Medicaid, with Medicare slated to move into first place. The first two are almost exclusively for the elderly (plus some of the disabled), and almost one-third of Medicaid expenditures already go to the elderly. The elderly vote at a high rate and they don't want cuts to these programs. The medical establishment in its various forms—which today is almost one-fifth of the economy—doesn't want those cuts either. The elderly will only increase in number, as will the medical establishment, so if we are not cutting these programs today why should we think that we will cut them ten or fifteen years down the road?

Most of all, the plausible cuts needed to balance the future budget are too large. To balance the budget right now through spending cuts, we'd basically have to come close to getting rid of Medicare, Medicaid, and Social Security altogether. And I'm talking about the complete elimination of those expenditures, not shifting them into somewhere or something else. There are plenty of plans to privatize some of these programs, and those are plans of varying quality, but they share the common property of shifting the costs of the programs rather than eliminating the costs of those programs.

I submit that the aggregate amount of aid given to the elderly, the needy, and other groups is unlikely to decline, whether we approve of that outcome or not. We may well increase some gaps in coverage, such as refusing to cover particular individuals (immigrants) or refusing to cover particular procedures (knee and back surgery, perhaps, until they become more effective). But the total expenditures on the health of the elderly will rise both in absolute and per capita terms, not fall. That is the general trend of Western societies since the late nineteenth century and no reformers, including Margaret Thatcher and Ronald Reagan, have really taken on entitlement spending through government and beaten it back. It's simply too popular.

Through all these programs, our altruism will remain intact or probably expand in terms of its absolute magnitude. That said, aid from the government will increasingly fall short of a growing set of demands, so unequal treatment will be more explicitly recognized as the norm. In percentage terms, relative to outstanding need and vociferous claims, the altruism of the public sector will have to fall. That will happen whether the left-wing progressives, the Tea Party, or some other group wins American elections. It's not about ideology; it's a question of making the numbers add up. Starting about ten years from now, the aging of the American population, combined with rising healthcare costs, will force on us some very radical fiscal changes.

I am forecasting a few particular changes, starting with the most obvious and ending with the least obvious:

1. We will raise taxes somewhat, especially on higher earners.
2. We will cut Medicaid for the poor (but not so much Medicaid for the elderly) by growing stingier with eligibility requirements and with reimbursement rates for Medicaid doctors, who will impose queuing on program beneficiaries.
3. The fiscal shortfall will come out of real wages as various cost burdens are shifted to workers through the terms of the employment relationship, including costly mandates.
4. The fiscal shortfall will come out of land rents; in other words, some costs of living will fall as people begin to live in cheaper housing.
5. We'll also pay off growing debt by spending less of our money on junk and wasteful consumption.

That's operating from a pretty simple theory of US politics that says "old people get their way." The corollary to that theory is "in the future they will get their way all the more." There will be more of them, and they are very often willing to vote on the issue of old-age benefits. By 2030, almost one-fifth of the United States population will be about as old as one-fifth of Florida residents today: sixty-five years old and up. Relatively few Florida politicians rise to the top, and achieve reelection, by calling for significant cuts to Social Security and Medicare.

It's a lot easier to cut Medicaid, and already a lot of the states are trying to escape paying their share for this program, which is jointly funded by federal and state governments. Medicaid is largely for the poor, and the poor vote at a much lower rate than do the elderly. The poor are also less influential and they tend to be less politically involved. They're too busy trying to make ends meet. State governments are trying to shift to lower Medicaid expenditures, not by making big announcements of cuts, but rather by changing the structure of Medicaid programs so that it is harder to actually collect on the benefits.

One way to think about the future cuts to Medicaid is to look at a typical state government budget. The big ticket items are education, roads, courts, and police. Medicaid expenses are pressing on those programs, and yet they are all more popular than spending more money on the health care of the poor.

Now let's look at the cuts to wages and the cuts to land rents and see how those might work.

When governments get into fiscal trouble or have to balance their budgets, they resort to mandates in the very broad sense of the word. For instance, if a government can't afford to pay for health insurance for everyone, it can mandate that businesses provide health insurance to (some of) their employees. That's exactly what the Obama healthcare reforms have done: specify that businesses above a certain size have to provide health insurance to their full-time employees. The self-employed and other groups have the individual mandate to buy health insurance with their own money, albeit combined with government subsidies.

The greater the value of the mandate, the less enthusiastic the business will be to hire more workers. Quite simply, mandates lower the demand for labor and create downward pressure on the general level of wages. And if you're told how to spend your own money, as the individual mandate does, this too lowers the value of your take-home pay.

The Obama healthcare legislation may or may not be good social policy, but it will make the labor market tougher for a lot of people and arguably the very prospect of the law already has done so. The Supreme Court, in upholding the individual mandate behind the bill, is implicitly recognizing the equilibrium I am outlining here. Given where America is starting from fiscally, mandates are difficult to avoid. It's also noteworthy that the one section of the legislation the court expressed reservations about was the Medicaid expansion, and over time that will come under increasing challenge.

Note that elderly voters—for selfish reasons—need not hate these mandates, because they are already covered by Medicare. Furthermore, without Medicare they would be buying health insurance in some manner anyway. Either the elderly are not working, or the mandates do not cut into their wages because the mandates do not apply to them. It's a good prediction to think that some kind of mandate will exist and survive, with the cost of that mandate being borne largely by labor markets and taken out of real wages paid to workers. I think of the future as a place where the ordinary worker will have more guaranteed access to health care on paper, not necessarily more actual access in practice, and less money to spend on other items of consumption.

How do we compensate for forthcoming fiscal problems with cheaper rents?

That too is pretty simple, although not entirely comforting. If your budget is tight, move to cheaper living quarters. That is the response you get, sooner or later, from people who are facing stagnant or lower wages, or from people who have had their government benefits cut or pared back.

This trend is already well underway.

One way to predict the future is to look at where people want to live right now. More concretely, we can look at which US states people move to as an indicator of what a lot of people really want. What I see is a big movement of Americans, and also immigrants, into Texas. For instance, during 2008–2009, the population of Texas increased, from migration alone, by more than 147,000 people.

Why is Texas so popular? For a long time the state has had one of America's highest murder rates and it has a high property crime rate. The weather is warm but it is not a calm warm-weather state, given the storms and tornados. Sometimes it is too warm, such as during the thirty-five straight days over a hundred degrees in July and August 2011 in the Dallas–Fort Worth area; worse yet, that wasn't a record. Texas is skimpy on welfare benefits and Medicaid coverage, and 27 percent of the state has no health insurance coverage. Texas has one of America's poorest performing educational systems, at least as measured by high school graduation rates, which in Texas are below 70 percent.

What Texas does have is very cheap housing and a decent record of job creation (you don't have to credit this to any particular Texas governor, any more than you should blame governors for the high murder rate). In other words, if you live in Texas, your locale will offer C-grade public services but you may have more cash in your pocket than if you lived somewhere else. You have a better chance of finding a job and will surely find cheaper housing. If you can adjust to the humidity, so much the better.

The difference in home prices is striking. For instance, a typical home in Brooklyn costs more than half a million dollars, and 85 percent of these dwellings are apartments and condos rather than stand-alone homes. They don't always have impressive sinks and air conditioning fixtures. In Houston the typical home costs $130,100, is likely a stand-alone, and stands a good chance of being newer.

The cheap housing doesn't just come from Texas's having a lot of land; there is another factor, namely that zoning in Texas is relatively weak. For instance, Houston doesn't have traditional zoning. You might find an office tower, a used-record store, and a whorehouse all right next to your home. Houstonians live with that, and since home prices are reasonable the relatively wealthy can insulate themselves from the less pleasant consequences of mixed-use neighborhoods. In any case, the absence of zoning makes the homes cheaper. I don't expect that trend to spread to all of America because suburban homeowner associations are politically powerful. Zoning may become relaxed in more parts of the country, but in the meantime people are voting with their feet and moving to Texas.

One lesson is that many other parts of America should have weaker zoning and cheaper land prices. Matthew Yglesias has written a good book, *The Rent Is Too Damn High*, arguing this point. I'll return to that, but, whether or not we make other real estate reforms across the country, there is a more fundamental, apparently obvious, and yet still underappreciated lesson: *People really like extra cash in their pocket.* They like that cash in their pocket more than our politicians wish were the case. You might think this desire is noble, à la Ayn Rand, or you might think it is selfishly unethical.

In any case, I'd like to explore what this love for the "filthy lucre"—which isn't going away—means for our future.

Since there is considerable net in-migration to Texas, I conclude that a lot of Americans would rather have some more cash than better public services. The other states experiencing significant in-migration are in the South and the less expensive parts of the West. For the most part, those are affordable states with decent job creation records, subpar public services on the whole, and cheap housing. Not everyone wants that bundle, as you will see if you poll the wealthy upper-middle-class residents of Brookline, Massachusetts or my own neighborhood in northern Virginia. Nonetheless, on the whole, we as a nation are moving in that direction.

Which are the states with the highest-quality public services? On the basis of measured expenditures this would typically be California and states in the Northeast, but in general those regions are seeing outmigration. Expensive government at the level of the state isn't winning the competitive tests.

Have you seen those lists of the world's "most livable" cities? Vancouver and Zurich always do well, and any visitor knows how beautiful and convenient they are. Yet Vancouver (the city proper) has only about 603,000 people and Zurich about 372,000 people. You can add on the broader metropolitan areas but still they don't come close to São Paulo or Tokyo in terms of population. I say the cities at the top of those lists are not actually so livable at all. They are extremely livable for media elites, the cosmopolitan wealthy, and vacationers. When it comes to the ordinary Joe, Jacksonville, Florida—however much it lacks Alpine beauty and European sophistication—is more livable than most of the cities on those "most livable" lists. As of 2006, Jacksonville's population was 794,555. I call that proof.

Many Americans will end up living in areas with cheaper housing and lower-quality public services, if only to give themselves more cash in their pocket. Some of those areas might be a bit ugly to the eyes, again as a trade-off for lower costs. As cross-country moving proceeds, and changes what we are, the United States as a whole will end up looking more like Texas.

This trend will shape our world whether or not you find that the cheaper cities suit you personally.

How far will it go? No one can say for sure, but the fiscal limitations of the American government, combined with wage polarization, will test that question.

When I visit Latin America, I am struck by how many people there live cheaply. In Mexico for instance, I have met large numbers of people who live on less than $10,000 a year, or maybe even on less than $5,000 a year. They hardly qualify as well-off but they do have access to cheap food and very cheap housing. They cannot buy too many other things. They don't always have money to bring the kid to the doctor or to buy new clothes. Their lodging is satisfactory, if not spectacular, and of course the warmer weather helps.

Let's say you are an American retiree who receives $22,000 in yearly income, plus Medicare coverage. That income may sound low but in fact it is the typical income for a Medicare recipient, keeping in mind that Medicare doesn't actually cover all healthcare expenses. On the plus side, you should note that many of these people receive family assistance or hold valuable assets, but still the median net worth of the American elderly is about $137,349—only a little above $20,000 if you exclude home equity.

Would many people at that income level like the option of paying Mexican-level prices for lodging and food, and receiving Mexican-quality lodging and food in return? You may not wish to actually execute that swap, but the *option* of making that move is still of value. For one thing, a decline in the value of investments could reduce your yearly income to $22,000 or below, causing you to reconsider and move south. In any case, if I were an elderly person I probably would rather live in Mexico on $22,000 a year than make do on that sum in American suburbia.

Of course, most Americans don't want to move to Mexico. Still, they might look more favorably on such an option if it were available in the United States. That will create pressures to allow such options in this country, for the same reasons that many people are moving to Texas.

These pressures to seek out cheaper living are already strong, but income polarization will intensify them. As a significant segment of the American labor force earns much more, they will bid up prices in the most desirable living areas. It will become harder to live in the nice parts of Los Angeles or Orange County—or even in the less nice parts, such as the still-expensive Anaheim. The need to move to a much cheaper area will grow. Meanwhile, the internet makes it more possible for at least some people to work at a greater distance, or to chat with their grandkids over Skype.

Income polarization, by the way, will have some more severe consequences for financial net worth than people might expect at first. By the time people get old, they are often not living off their income but living off their wealth. For a given difference in lifetime income, between two groups of people, the eventual difference in wealth is usually much greater. The people with the higher incomes have saved more, started more businesses, avoided debt, and perhaps invested more wisely. They have received medical attention when needed and thus maintained their productivity better. In other words, initial income differences compound over time and wealth inequality in the United States (and elsewhere) is much greater than income inequality. That will give a further boost to the premium for living in a nice area. Once again, the less wealthy will be pushed out of the nicer living areas and this will be done by market prices, combined with some restrictive government zoning.

Think through the budget of a typical elderly individual living off $22,000 per year. Let's put that person in Knoxville, Tennessee, which I take to be a fairly typical part of this country, insofar as that phrase may be invoked at all. A one- or two-bedroom apartment in Knoxville, not the fancy parts, might cost in the range of $700 to $1,200 a month. That can easily run up to half of the overall budget, and it is likely the single largest expense of that person. Insofar as people are squeezed by circumstance, it is one expense that is potentially controllable.

What if someone proposed that in a few parts of the United States, in the warmer states, some city neighborhoods would be set aside for cheap living? We would build some "tiny homes" there; tiny homes might be about 400 square feet and cost in the range of $20,000 to $40,000. We would build some very modest dwellings there, as we used to build in the 1920s. We also would build some makeshift structures there, similar to the better dwellings you might find in a Rio de Janeiro *favela*. The quality of the water and electrical infrastructure might be low by American standards, though we could supplement the neighborhood with free municipal wireless (the future version of Marie Antoinette's famous alleged phrase will be "Let them watch internet!"). Hulu and other web-based TV services would replace more expensive cable connections for those residents. Then we would allow people to move there if they desired. In essence, we would be recreating a Mexico-like or

Brazil-like environment in part of the United States, although with some technological add-ons and most likely with greater safety.

Many people will be horrified at this thought. How dare you propose we stuff our elderly into shantytowns? Maybe they are right to be upset, although recall that no one is being *forced* to live in these places. Some people might prefer to live there. I might prefer to live there if my income were low enough. Don't confuse railing against the causes of low incomes with railing against how people adjust to low incomes; the latter adjustments are usually beneficial, even if you think the low incomes are unnecessary or unjust in the first place. (We might try to fix the causes of the low incomes—and on that, see the education chapter—rather than trying to outlaw all of the symptoms of lower incomes.)

But imagine no one confronts any explicit decision to set up shantytowns. Through the normal process of political evolution and indeed through sheer randomness, some parts of the country become more run-down than others. Their public services deteriorate and their real estate prices fall accordingly. No one is willing to spend the money on fixing up these places, whether through the private sector or the public sector. More poor people, including more poor elderly, go live there. There is no free municipal wireless. We evolve our way into some partial shantytowns, although without ever calling them that. The end result isn't so different from the deliberate shantytowns already discussed.

We already see experiments with blending much lower living costs into leading cities. El Paso is America's twenty-third largest city (using data from 2000) and it would jump to fifth largest if we combined it with the population of its sister city, Ciudad Juarez, right across the Mexican border. You can think of it as one consolidated city with a very large attached shantytown. Indeed, this shanty is essential to the success of El Paso. El Paso lives off of the manufacturing across the border, as it lost its own manufacturing base some time ago and it has substandard levels of education. The city also benefits from an army base, from border-related law enforcement efforts, and from the drug trade. Howard Campbell, an anthropologist, notes that El Paso is parasitic off of Juarez rather than vice versa. El Paso has flourished by hooking up with an adjacent neighbor with much lower rent and much lower quality infrastructure. Despite all the problems that can cross the border, and despite Juarez being one of the world's drug-cartel and murder capitals, few people in El Paso actually wish for Juarez to go away.

Another successful low-rent city is Berlin, which is now Europe's capital for contemporary culture and its most exciting bohemian youth city. Low rents in Berlin stem mainly not from deteriorating infrastructure (although there is some of that in the eastern part of the city), but rather from previous overbuilding relative to current needs. Berlin was once Germany's leading economic powerhouse, but the Nazis and the Communists wrecked the place. It is a city built to be a business capital but is no longer anything close to a business capital. That means overbuilding and that means lots of things are remarkably cheap, at least by European standards. It is easy to rent an acceptable apartment in a non-peripheral part of Berlin, not too far from a subway or streetcar stop, for a few hundred dollars a month. Food, too, is much cheaper than in the rest of Western Europe—cheaper than in most of the rest of Germany even. There are many thousands of people in Berlin simply living, on low rent, to "get by." It's the ultimate slacker city.

The most extreme low-rent move is to go "off the grid." For all the technological progress we have seen, a growing number of Americans are disconnecting from traditional water and electricity hookups and making their own way, often in owner-built homes, micro-homes, trailer parks, floating boats, or less elegantly in tent cities, as we find scattered around the United States, including in Portland, Seattle, and Los Angeles. Some of these options are gruesome, but many people are doing it by choice. New technologies, such as powerful local generators and solar power, are making it easier to strike out on one's own.

I'm saying that lower land prices are one big way many Americans are going to adjust to uneven patterns of wage progress and to the forthcoming fiscal squeeze. We're going to get to lower land prices one way or another. Not Manhattan, not West LA, not Fairfax County, Virginia, and not the whole country, but some parts of it. Some version of Texas—and then some—is the future for a lot of us. Lower prices—enjoyed selectively by aid recipients—raise the real value of aid and make Social Security checks go further. That's one way of finding some "give" in our budget dilemma. People will respond to stagnant or shrinking entitlements by moving to cheaper areas.

Kurt Vonnegut, in his 1952 novel *Player Piano*, also saw mechanization and labor market evolution as bringing more residential segregation. The novel starts like this:

> Illium, New York, is divided into three parts.
> In the northwest are the managers and the engineers and civil servants and a few professional people; in the northeast are the machines; and in the south, across the Iroquois River, is the area locally known as Homestead, where almost all of the people live.

There is one final way we will adjust to uneven wage patterns and that is with our tastes. Many of society's lower earners will re-shape their tastes—will *have to* reshape their tastes—toward cheaper desires. Caviar is an expensive desire and Goya canned beans is a relatively cheap desire. Don't scoff at the beans: With an income above the national average, I receive more pleasure from the beans, which I cook with freshly ground cumin and rehydrated, pureed chilies. Good tacos and quesadillas and tamales are cheap too, and that is one reason why they are eaten so frequently in low-income countries.

The bad news is that there is a lot of waste in American consumption—massive amounts of waste, in fact. Everyone has their favorite story about what the other guy spends his money on and could do without. But also the good news, oddly enough, is that there is a lot of waste in American consumption. Citizens faced with financial pressures will shift into cheaper consumption, and a lot of them will do so without losing very much happiness or value, precisely because there is already so much waste in what they buy.

You can think of the budgetary pressures—and machine intelligence—as pulling us all in the same direction, namely getting rid of that wasteful consumption. Imagine seeing your consumer budget get tighter every year. Also imagine getting a weekly report from your smart phone: "These are the purchases that really got your heart racing," followed by a list. You would then receive "These are the purchases you set aside and never used again, or that maybe didn't even give you a brief smile," followed by another list. You won't have to be a genius to put together the two developments and adjust your spending habits accordingly.

This process of economizing won't always go so well when it comes to poor women. A recent Pew Research Center study examined exactly who in modern America falls out of the middle class, and it found that women who are divorced, widowed, or separated are an especially vulnerable group. And children don't help single mothers' incomes. Taking care of one's children can be thought of as a very expensive preference but it is a preference that, for a lot of people, is not going away. Younger women in the lower end of the income distribution will probably be some of the biggest losers, especially if they have a strong "baby lust" that induces or compels them to have lots of kids early in life. Many of these women will also find it harder to move to cheaper areas with lower-quality infrastructure because they may still desire good schools for their children, especially if those kids are not self-starting learners from the internet. To top off all these problems, the desire for cheaper preferences and lifestyles may induce more lower-income men to abandon their children or at least to scale back financial support, a development that is extensively cataloged in conservative critic Charles Murray's book *Coming Apart*.

Just as some poorer people will do without fancy infrastructure, so will others do without advanced health care. Since we won't be willing to pay for full-benefit Medicare and Medicaid for everyone who will need it, some people will see cut benefits or rationed access to doctors. Our political system will try to construct that rationing so that voters blame the doctors rather than the politicians, but one way or another rationing will increase. Imagine many more millions of people wishing to see a doctor and having to wait weeks or months to do so.

The disparities of healthcare access, across classes of income and education, will increase. The world will look much more unfair and much less equal and indeed it will be.

That sounds pretty bad and maybe it is. There will, however, be some compensating factors. The first, as I've already noted, is that more people will get more aid than ever before, even though the quality of that care—and how long many will have to wait to get it—is anyone's guess.

Second, if you sit around and talk with health policy wonks, it is remarkable how skeptical they can be about the value of marginal increments of health care. A lot of factors go into determining actual health outcomes, including exercise, diet, taking care of one's self, perhaps attitude, demographic status, and so on. If you pose the question "What percent of health outcomes is determined by health care?" you'll get some pretty modest estimates. I've heard figures of no more than 10 to 15 percent. Maybe that's undercounting and maybe it's 30 or 40 percent; we don't seem to know. In any case, that suggests there is room for some people to win back those healthcare losses by taking better care of themselves.

Let's say you were eligible for Medicaid but the government cuts the value of your healthcare benefit by 40 percent. Depending on your age and condition, you can make up some or all of that and maybe then some. Exercise more. Give up junk food. If you are a man, you will probably live longer if you marry. So pull out that old photo and get your fifty-three-year-old body on Match.com.

Not everyone will respond in this way. We'll end up with a society where the people with decent self-control win back a lot of the lost health gains by better behavior. The people who don't have good self-control will lose out much more. They'll lose a chunk of their health care and they won't respond by getting on that exercise bike.

Personal qualities of character such as self-motivation and conscientiousness will reap a lot of gains in the new world to come. We can already see this in the numbers. The individuals falling out of the middle class are more likely to be divorced, to have low levels of formal education, to have low test scores, and to have a history of drug use.

The Politics of the Future

With wage polarization, and a lot of the elderly and poor living in lower-rent areas, what will our future politics look like?

In the wake of Occupy Wall Street, the Tea Party movement, and growing income inequality, a lot of commentators are predicting an America torn by protest and maybe political violence. I do think we'll see some outbursts of trouble, but the longer-run picture will be fairly calm and indeed downright orderly. I expect a society that will be more conservative, both politically and in the more literal sense of that term.

For all the prognostications about the American future, the most important single fact, and the easiest to predict, is simply that we will be a lot older. That will make us more conservative, in this case referring to the literal rather than political sense of that term. Revolutions and protests are the endeavors of young hotheads, not sage (or tired) sixty-four-year-olds. The societies with lots of un-married young men are the most vulnerable to sudden revolutions and major political changes. Large parts of the Arab world fit this designation, and thus we have seen the Arab Spring, but we Americans are moving along a different path.

As I've mentioned, right now about 19 percent of Florida is over the age of sixty-five. By 2030, 19 percent of the United States will be over sixty-five years of age; in other words, we'll be like Florida in terms of age structure. We'll then get older yet. We will be more cautious and more reluctant to change things very rapidly. We also may be less *able* to change things very rapidly. People will be set in their ways, aid to the elderly will soak up a large part of the government budget, and wealthy earners will be the dominant economic, social, and political influence.

Commentators often suggest that wage polarization will mean the end of liberalism—meaning a broadly tolerant society with lots of liberty and a protected personal sphere—or the end of democracy. We can imagine the lower-wage individuals toppling the proverbial Bastille and taking away the goodies of the higher earners. These are tempting conclusions, but there's not much evidence to support them. Societies have a strong status quo bias, particularly if they have high status relative to other parts of the world. Hardly anyone need feel bad about being an American, and twenty years from now that is unlikely to be much different. For the foreseeable future, America will still claim a place as world leader or, at the very worst, part of a bipolar global rivalry with China. A strong rivalry with China will, if anything, likely cement feelings of American nationalism and strengthen an orderly status quo, just as the Cold War with the Soviet Union did. Whatever you may think of this future from your 2013 vantage point, people will look around and still see that America is one of the nicest places in the world. That's hardly a recipe for revolutionary fervor.

If you're trying to measure the scope or potential for social disorder, look at the rate of crime. In the United States crime rates have been falling for decades and in recent times they have surprised researchers by falling even faster than expected. Yet over those same decades income and wealth inequality have been rising significantly in the United States. It seems that, whether we like it or not, increasing inequality and growing domestic peace are compatible. Very often I read warnings about how income inequality will lead to a society where the poor take by force what they cannot earn in the marketplace. Yet these predictions run aground on the simplest of empirical tests, namely crime rates.

The last time American society had a lot of demonstrations and riots was in the 1960s and early 1970s. There was a war with a lot of domestic casualties, a draft, major race riots, the total or partial occupation of major university campuses, and a young, volatile population. Those days are far away and Occupy Wall Street is remarkably tame in comparison. It's worth noting that the 1960s were a kind of golden age for income equality, American manufacturing, rising incomes, and a general sense of American prosperity. In fact, it's the last time that median incomes for American households rose consistently on a sustained basis. There was plenty of talk of oppression and unfairness, but today's "top 1 percent" was not a major concept and investment bankers might have earned no more than $100,000 a year, admittedly in a world of lower prices but still that is no comparison to today's compensation disparity.

A lot of commentators, most of all from the progressive Left, object strenuously to rising wealth and income inequality. Even if they are correct in their moral stance, they too quickly conclude that rising inequality has to cause other bad results, such as revolution, expropriation, or a breakdown in social order. That does not follow, and I sometimes wonder if it isn't an internal psychological mechanism operating in some of these commentators, almost as if they were wishing for the wealthy to be punished for their sins. My skepticism toward these hypotheses of disorder is not just driven by my recognition of the general aging of the population or the falling crime rates. There are many other historical periods, including medieval times, where inequality is high, upward mobility is fairly low, and the social order is fairly stable, even if we as moderns find some aspects of that order objectionable.

I wonder if this "threat of revolution" argument isn't a substitute for actually making a good case for a feasible reform. I've very often heard commentators from the Left suggesting that if we don't "do something" about income inequality, citizens will take matters into their own hands. There is a vague insinuation of a threat of violence, yet without any endorsement (or condemnation) of that violence. The commentator or writer doesn't want to suggest that the violence is in order, yet still wants the rhetorical force of having that violence on his or her side of the argument, as a kind of cosmic punishment for the objectionable inequality. A simple question: Would it occur to such writers to fear such a violent outbreak and then preach to their fellow citizens to therefore be less concerned with inequality? I don't think so, and I take that as a sign that the predictions of violence tell us more about the predictors than about the likely course of future American society. Inequality can have some bad consequences, and we are likely to experience some of those consequences, but these predictions of durable and significant social unrest are some of the least thought-out and least well-supported arguments with wide currency.

The better guess is that Americans will become more conservative, now returning to both the political and literal senses of that word. They will become more enamored of low or falling taxes, whether or not such tax rates prove possible to maintain. They will want to be promised something for nothing. They will look more toward local communities and tight local bonds, to protect themselves against economic risks. Unlike the predicted breakdown in social order, these trends are already significant and observable in today's America.

We will at the same time see a lot of ugly discourse, a lot of polarization in Congress, and a lot of distasteful political warfare. The breakdown in mainstream media and the decentralization of debate over the internet will bring more partisanship and more name calling. It may resemble the name-calling Jacksonian newspapers of the nineteenth century and the incessant scandal mongering found throughout most of American political history. It is the relatively bland mainstream media of the 1950s and '60s that represents the outlier, and the discourse has already turned back closer to historical norms.

Trends in campaign finance reform have made parties, politicians, and campaigns more dependent on the political base and on relatively extreme donors. That has helped make Congress dysfunctional, but we shouldn't confuse the rancor of those debates with the underlying mood of the American voter. Most American voters are fairly moderate, disillusioned with both political parties, and looking for someone who can fill the proverbial niche of "getting something done," or "unifying the nation." Those are not the kind of attitudes that make for a revolutionary future, and they are too ordinary to fill up the time on TV, talk radio, or the political blogs. The most likely winning political coalition of the future is a rather unglamorous mix of such moderates, a lot of the elderly, and one chunk of the elite, mostly oriented toward the status quo.

It's again worth seeing what is happening, politically speaking, in the parts of the United States with relatively stagnant incomes. Political conservatism is strongest in the least well-off, least educated, most blue collar, and most economically hard-hit states. If you doubt it, know that as of 2011, the most politically conservative states are, as measured by self-identification, Mississippi, Idaho, Alabama, Wyoming, Utah, Arkansas, South Carolina, North Dakota, Louisiana, and South Dakota. As Richard Florida puts it, "Conservatism, more and more, is the ideology of the economically left behind."

Those states have become outposts of Tea Party support. Their electorates are not out there leading the charge for higher rates of progressive taxation or trying to revive the memory of George Mc-Govern. The most liberal areas tend to be urban or suburban, with lots of high-earning professionals. My own residence—in Fairfax County, Virginia—was strongly conservative in the early 1980s when I first lived there. It voted reliably conservative and Republican or it insisted on quite conservative Democrats. It was well-off but not thought of as a bastion of riches. Run-down neighborhoods still were common. Circa 2012, Fairfax County is now in per capita terms the wealthiest county in the United States. It broke cleanly for Obama in the 2008 and 2012 elections and it is somewhat more Democratic than Republican in terms of party support, although the region still is not conducive to the more radical sides of the Left such as you might find in, say, Berkeley, California.

Or look at where Occupy Wall Street has been strong as a movement. It holds great appeal for well-educated young people from the upper middle class, especially if they are underappreciated liberal arts majors who do not have the option of stepping into the highest-paying or most upwardly mobile jobs. It is not a broader American phenomenon that is catching fire on the docks of Elizabeth, New Jersey, or in the ailing Appalachian regions of Ohio or with religious homeschoolers in Idaho.

If we extrapolate these trends into the future, we can expect the higher earners to identify with the values embraced by today's moderate Democrats. They will believe in progress, diversity, and social justice, although they may not be huge fans of radically progressive taxation. Some of them will be "small L libertarians," but those libertarians will like the same jokes and TV shows as the moderate Democrats among the high earners. The lower earners will be split into two groups, the more extreme conservatives versus the individuals who receive transfers from the social welfare programs supported by the moderate Democrats. I'm not suggesting that they will cynically vote to line their pockets, rather that the moderate Democrats will offer a worldview that embraces those individuals and offers them higher status and respect, thus winning their political loyalties. The more extreme conservatives will embrace religion and nationalism to a higher degree. This is already a pretty fair description of the American political spectrum and it is consistent with a world of growing income and labor market polarization.

If you think about it, we really shouldn't expect rising income and wealth inequality to lead to revolution and revolt. That is for a very simple psychological reason: Most envy is local. At least in the United States, most economic resentment is not directed toward billionaires or high-roller financiers—not even corrupt ones. It's directed at the guy down the hall who got a bigger raise. It's directed at the husband of your wife's sister, because he earns 20 percent more than you do. It's directed at the people you went to high school with. And that's why a lot of people aren't so bothered by income or wealth inequality at the macro level: Most of us don't compare ourselves to billionaires. Gore Vidal put it honestly: "Whenever a friend succeeds, a little something in me dies." Right now the biggest medium for envy in the United States is probably Facebook, not the yachting marinas or the rather popular television shows about the lifestyles of the rich and famous.

Sometimes I wonder why so many relatively well-off intellectuals lead the egalitarian charge against the privileges of the wealthy.

One group has the status currency of money and the other has the status currency of intellect, so might they be competing for overall social regard? And in that competition, at least in the United States, the status currency of intellect is not winning out. Perhaps for that reason the high status of the wealthy in America, or for that matter the high status of celebrities, bothers our intellectual class most. That intellectual class, however, is small in number, so growing income inequality won't by itself lead to a political revolution along the lines many intellectuals have imagined.

The American polity is unlikely to collapse, but we'll all look back on the immediate postwar era as a very special time. Our future will bring more wealthy people than ever before, but also more poor people, including people who do not always have access to basic public services. Rather than balancing our budget with higher taxes or lower benefits, we will allow the real wages of many workers to fall and thus we will allow the creation of a new underclass. We won't really see how we could stop that. Yet it will be an oddly peaceful time, with the general aging of American society and the proliferation of many sources of cheap fun. We might even look ahead to a time when the cheap or free fun is so plentiful that it will feel a bit like Karl Marx's communist utopia, albeit brought on by capitalism. That is the real light at the end of the tunnel. Such a development, however, will take longer than I am considering in the time frame of this book.

In the meantime, get ready. The basic look of our lives, and the surrounding physical environment, hasn't been revolutionized all that much in forty to fifty years—just try viewing a TV show from the 1970s and the world will seem quite familiar. That's about to change. It is frightening, but it is exciting too.

It might be called the age of genius machines, and it will be the people who work with them that will rise. One day soon we will look back and see that we produced two nations, a fantastically successful nation, working in the technologically dynamic sectors, and everyone else. Average is over.

Notes

For the opening quotation, see D.T. Max, "The Prince's Gambit: A Chess Star Emerges for the Post-Computer Age," *The New Yorker*, March 21, 2011.

Chapter 1: Work and Wages in iWorld

For the figures on wages of college graduates, see Heidi Shierholz, Natalie Sabadish, and Hilary Wething, "The Class of 2012: Labor Market for Young Graduates Remains Grim," Economic Policy Institute, May 3, 2012, http://www.epi.org /publication/bp340-labor-market-young-graduates/. For differing and indeed more pessimistic estimates, see Michael Mandel, "The State of Young College Grads 2011," Mandel on Innovation and Growth, October 1, 2011, http://innovationandgrowth. wordpress.com/2011/10/01/the-state-of-young-college-grads-2011/, and also "Bad Decade for Male College Grads," September 25, 2011, http://innovationandgrowth. wordpress.com/2011/09/25/bad-decade-for-male-college-grads/. Mandel's work has been updated by Diana G. Carew, "Young College Grads: Real Earnings Fell in 2011," blog: *The Progressive Fix*, September 20, 2012, http://www.progressivepolicy. org/2012/09/young-college-grads-real-earnings-fell-in-2011/. Again, these data are for the individuals not going on for graduate degrees. The underlying raw data come from the Census Bureau, Table P-32, Educational Attainment, currently online at http://www.census.gov/hhes/www/income/data/historical/people/.

On computing speeds for the iPhone, see Charles Arthur, "How the Smartphone is killing the PC," *The Guardian*, June 5, 2011.

In addition to the books cited on robots and machine intelligence, see also some of Paul Krugman's 2012 and 2013 blog posts on robots, related blog posts by Nick Rowe at Worthwhile Canadian Initiative, and Jeffrey D. Sachs and Laurence J. Kotlikoff, "Smart Machines and Long-Term Misery," National Bureau of Economic Research, Working Paper 18629, December 2012. Izabella Kaminska has assembled many of the relevant blog posts here: http://theleisuresociety.tumblr.com/post/39057729530 /the-tech-debate-blasts-off-a-linkfest. See also Krugman's earlier essay "Technology's Revenge," available on The Unofficial Paul Krugman Archive, http://www .pkarchive.org/economy/TechnologyRevenge.html, originally published 1994.

On South Korean prison wardens, see "Robotic Prison Wardens to Patrol South Korean Prison," BBC News, November 25, 2011.

The joke about the factory and the dog is cited in Adam Davidson, "Making it in America," *The Atlantic*, January/February 2012.

On journalism, see Steve Lohr, "In Case You Wondered, a Real Human Being Wrote This Column," *The New York Times*, September 10, 2011. On grading software, see Jim Giles, "AI Makes the Grade," *New Scientist*, September 4, 2011, p. 22.

On dating algorithms, see for instance David Gelles, "Inside Match.com," *Financial Times*, July 29, 2011.

On Santa Cruz and crime, see G.O. Mohler, M.B. Short, P.J. Brantingham, F.P. Schoenberg, G.E. Tita, "Self-Exciting Point Process Modeling of Crime," *Journal of the American Statistical Association*, March 2011, 106(493): 100–108, doi:10.1198/jasa.2011.ap09546, and see also James Vlahos, "The Department of Pre-Crime," *Scientific American*, January 2012. On the TSA, see Joseph A. Bernstein, "Big Idea: Seeing Crime Before It Happens," *Discover*, December 2011. Most generally, see Erica Goode, "Sending the Police Before There's a Crime," *The New York Times*, August 15, 2011.

On predictions of pregnancy, see Charles Duhigg, "Psst, You in Aisle 5," *The New York Times Sunday Magazine*, February 19, 2012.

On headgear and the following of vision, see Adam Piore, "Ailment: Too Much Information, Cure: Mind-Reading Machines," *Discover*, October 2011, p. 38. On being tracked by sensors and subject to retina scans, see Martin Lindstrom, "Shopping Carts Will Track Consumers' Every Move," *Harvard Business Review* blog, December 9, 2011. See also Ashley Lutz and Matt Townsend, "Big Brother Is Watching You Shop," *Bloomberg Businessweek*, December 15, 2011. On retinas, see Emily Glazer, "The Eyes Have It: Marketers Now Track Shoppers' Retinas," *The Wall Street Journal*, July 12, 2012.

On rooting out fake reviews, see David Streitfeld, "In a Race to Out-Rave, 5-Star Web Reviews Go for $5," *The New York Times*, August 19, 2011. On discerning cheating in online dating profiles, see Catalina L. Toma and Jeffrey T. Hancock, "What

Lies Beneath: The Linguistic Traces of Deception in Online Dating Profiles," *Journal of Communication*, February 2012, 62(1): 78–97.

On lie detection, see Anne Eisenberg, "Software That Listens for Lies," *The New York Times*, December 3, 2011.

Chapter 2: The Big Earners and the Big Losers

One good take on distributional issues is Jérémie Cohen-Setton, "Blogs Review: Robots, Capital-Biased Technological Change and Inequality," Bruegel's Improving Economic Policy blog, December 10, 2012, http://www.bruegel.org/nc/blog/detail /article/958-blogs-review-robots-capital-biased-technological-change-and -inequality/.

On drone employment, see David S. Cloud, "Civilian Contractors Playing Key Roles in U.S. Drone Operations," the *Los Angeles Times*, December 29, 2011.

For some sources on income inequality, see for instance Steven N. Kaplan and Joshua Rauh, "Wall Street and Main Street: What Contributes to the Rise in the Highest Incomes?", *The Review of Financial Studies*, 2010, 23(3): 1004–1050; Howard Wial, "Where the 1% Live," Atlantic Cities blog, October 31, 2011; and Chris Forman, Avi Goldfarb, and Shane Greenstein, "The Internet and Local Wages: A Puzzle," *American Economic Review*, Febraury 2012, 102(1): 556–75. On executives, managers, and supervisors, see Jon Bakija, Adam Cole, and Bradley T. Heim, "Jobs and Income Growth of Top Earners and the Causes of Changing Income Inequality: Evidence from U.S. Tax Return Data," an unpublished working paper.

For the anecdotes about being "acqhired," see Miguel Helft, "For Buyers of Web Start-Ups, Quest to Corral Young Talent," *The New York Times*, May 17, 2011.

On job categories that have improved, see Michael Mandel, "Three Industries That Have Continued to Add Jobs," Mandel on Innovation and Growth, September 2, 2011, http://innovationandgrowth.wordpress.com/2011/09/02/three-industries -that-continue-to-add-jobs/. On STEM jobs and science degrees, see Anthony P. Carnevale, Nicole Smith, and Michelle Melton, "STEM," Georgetown University Center on Education and the Workforce, http://www9.georgetown.edu/grad/gppi /hpi/cew/pdfs/stem-complete.pdf.

For one look at the costs of bad workers, see Will Felps, Terence R. Mitchell, and Eliza Byington, "How, When, and Why Bad Apples Spoil the Barrel: Negative Group Members and Dysfunctional Groups," *Research in Organizational Behavior*, 2006, 27: 175–222. The most comprehensive study on how higher-quality workers are clustering together to greater degree is David Card, Jörg Heining, and Patrick Kline, "Workplace Heterogeneity and the Rise of West German Wage Inequality," National Bureau of Economic Research, Working Paper 18522, November 2012. On the underlying theory, see Michael Kremer, "The O-Ring Theory of Economic Development," *The Quarterly Journal of Economics*, August 1993, 108(3): 551–75.

On women being more conscientiousness, see for instance David P. Schmitt, Anu Realo, Martin Voracek, and Jüri Allik, "Why Can't a Man Be More Like a Woman? Sex Differences in Big Five Personality Traits Across 55 Cultures," *Journal of Personality and Social Psychology*, 2008, 94(1): 168–82, doi:10.1037/0022-3514.94.1.16. An important paper on related topics is Paul Beaudry and Ethan Lewis, "Do Male-Female Wage Differentials Reflect Differences in the Return to Skill? Cross-city Evidence from 1980–2000," National Bureau of Economic Research, Working Paper 18159, June 2012. On female success in working in teams, see Jeffrey A. Flory, Andreas Leibbrandt, and John A. List, "Do Competitive Work Places Deter Female Workers? A Large-Scale Natural Field Experiment on Gender Differences in Job-Entry Decisions," National Bureau of Economic Research, Working Paper 16546, November 2010. On the importance of conscientiousness in the workplace and for earnings, see for instance Murray R. Barrick and Michael K. Mount, "The Big Five Personality Dimensions and Job Performance: A Meta-Analysis," *Personnel Psychology*, 1991, 44(1): 1–26; Ellen K. Nyhus and Empar Pons, "The Effects of Personality on Earnings," *Journal of Economic Psychology*, 2005, 26(3): 363–84; and Daniel Spurk and Andrea E. Abele, "Who Earns More and Why? A Multiple Mediation Model from Personality to Salary," *Journal of Business and Psychology*, 2011, 26(1): 87–103. For more general writings on conscientiousness, see Brent W. Roberts, Carl Lejuez, Robert F. Krueger, Jessica M. Richards, and Patrick L. Hill, "What Is Conscientiousness and How Can It Be Assessed?", *Developmental Psychology*, online first publication, December 31, 2012, and also Angela L. Duckworth, David Weir, Eli Tsukayama, and David Kwok, "Who Does Well in Life? Conscientious Adults Excel in Both Objective and Subjective Success," *Frontiers in Personality Science and Individual Differences*, online first publication, September 28, 2012.

For the male unemployment rate for 2012, see the Bureau of Labor Statistics, http://www.bls.gov/news.release/empsit.t10.htm.

On teen employment falling, see Christopher L. Smith, "Polarization, Immigration, Education: What's Behind the Dramatic Decline in Youth Employment?", Finance and Economics Discussion Series, Federal Reserve Board, October 2011.

For the cited poll of managers, and which skills are needed, see Timothy F. Bresnahan, Erik Brynjolfsson, and Lorin M. Hitt, "Information Technology, Workplace Organization, and the Demand for Skilled Labor: Firm-Level Evidence, National Bureau of Economic Research, Working Paper 7136, May 1999.

The Henry Mayhew quotation is from *London Labour and the London Poor* (Oxford: Oxford University Press, 2012), 141, with the original being published in installments in the 1840s.

The Google questions are from Nicholas Carlson, "15 Google Interview Questions That Will Make You Feel Stupid," *Business Insider*, November 4, 2009.

On fire chiefs and master's degrees, see Paul Fain, "Advanced Degrees for Fire Chiefs," *Inside Higher Ed*, October 27, 2011, http://www.insidehighered.com/news/2011/10/27/college-degrees-increasingly-help-firefighters-get-ahead#ixzz1f0qkakYi. On rising degree requirements more generally, see Catherine Rampell, "Degree Inflation? Jobs that Newly Require B.A.'s," *The New York Times* Economix blog, December 4, 2012.

That jobs lost during the recession came from the middle, see "The Good Jobs Deficit," National Employment Law Project, July 2011, http://www.nelp.org/page/-/Final%20occupations%20report%207-25-11.pdf?nocdn=1, also discussed in Steven Greenhouse, "Where the Job Growth Is: At the Low End," *The New York Times* Economix blog, July 27, 2011. On the longer-term trends, see David Autor, "The Polarization of Job Opportunities in the U.S. Labor Market: Implications for Employment and Earnings," *Community Investments*, Fall 2011, 23(2): 11–41. See also the important piece by David H. Autor, Lawrence F. Katz, and Alan B. Krueger, "Computing Inequality: Have Computers Changed the Labor Market?", *The Quarterly Journal of Economics*, November 1998, 113(4): 1169–1213.

For a look at some Census Bureau data, see Carmen DeNavas-Walt, Bernadette D. Proctor, and Jessica C. Smith, "Income, Poverty, and Health Insurance Coverage in the United States: 2011, Current Population Reports," September 2012, http://www.census.gov/prod/2012pubs/p60-243.pdf. See also the discussion in Robert Pear, "Recession Officially Over, U.S. Incomes Kept Falling," *The New York Times*, October 9, 2011.

On the new and cheaper manufacturing jobs, see Louis Uchitelle, "Factory Jobs Gain, but Wages Retreat," *The New York Times*, December 29, 2011.

On the decline in labor's share, see Francisco Rodriguez and Arjun Jayadev, "The Declining Labor Share of Income," United Nations Development Programme Human Development Reports Research Paper, November 2010, http://hdr.undp.org/en/reports/global/hdr2010/papers/HDRP_2010_36.pdf, and Florence Jaumotte and Irina Tytell, "How Has the Globalization of Labor Affected the Labor Income Share in Advanced Countries?", IMF Working Paper, 2007. In particular see this latter piece on rising income shares for skilled labor. See also Peter Orszag, "As Kaldor's Facts Fall, Occupy Wall Street Rises," Bloomberg.com, October 18, 2011. A good survey is John van Reenen, "Wage Inequality, Technology and Trade: 21st Century Evidence," Centre for Economic Performance, London School of Economics and Political Science, May 2011.

On labor market polarization in Europe, see Maarten Goos, Alan Manning, and Anna Salomons, "Explaining Job Polarization in Europe: The Roles of Technology, Globalization, and Institutions," CEP Discussion Paper 1026, Centre for Economic Performance, November 2010.

On the importance of wage gains for advanced-degree holders, see David Wessel, "Only Advanced-Degree Holders See Wage Gains," Real Time Economics blog, *The Wall Street Journal*, September 19, 2011, based on work by Matthew Slaughter and related to Census data, "Income, Poverty, and Health Insurance Coverage in the

United States: 2010, Current Population Reports," September 2011, http://www
.census.gov/prod/2011pubs/p60-239.pdf.

Chapter 3: Why Are So Many People Out of Work?

On these and other factors behind labor force participation, see Willem Van Zand-
weghe, "Interpreting the Recent Decline in Labor Force Participation," Federal
Reserve Bank of Kansas City, 2012.

On Belle, see Joe Condon and Ken Thompson, "Belle Chess Hardware," reprinted
in, *Computer Chess Compendium* (New York: Ishi Press International, 2009),
286–92, David Levy, editor.

On nonhuman DJs, see John Roach, "Non-Human DJ Gets Radio Gig," NBC News,
www.today.com/tech/non-human-dj-gets-radio-gig-121286.

On various points concerning unemployment and labor force participation, see
David Wessel, "What's Wrong with America's Job Engine," *The Wall Street Journal*,
July 27, 2011. See also Brad Plumer, "The Incredible Shrinking Labor Force," *The
Washington Post*, May 4, 2012. The statistics themselves come from the Bureau of
Labor Statistics; see for instance http://www.bls.gov/web/empsit/cpseea03.pdf.

On unemployment and men not working, see David Leonhardt, "Men, Unemployment,
and Disability," *The New York Times* Economix blog, April 8, 2011. On younger men, see
"Employment Participation Rate Shows a Troubling Trend," *Sober Look,* February 6,
2012, http://soberlook.com/2012/02/employment-participation-rate-shows.html.

On disability, see David H. Autor, "The Unsustainable Rise of the Disability Rolls
in the United States: Causes, Consequences, and Policy Options," National Bureau
of Economic Research, Working Paper 17697, December 2011. For some very recent
disability estimates, see the Social Security Administration, http://www.social
security.gov/cgi-bin/currentpay.cgi.

For Scott Winship on male median wages, see Scott Winship, "Men's Earnings Have
NOT Declined by 28 Percent Since 1969!" March 29, 2011, http://www.scottwinship.com
/1/post/2011/03/mens-earnings-have-not-declined-by-28-percent-since-1969.html. For
the original source he is criticizing, see Michael Greenstone and Adam Looney, "Trends:
Reduced Earnings for Men in America," Brookings Institution, November 2011.

On the special weaknesses of the labor market, this time around, see Menzie Chinn,
"Takes from the GDP Revisions," *Econbrowser*, July 31, 2011, http://www.econ
browser.com/archives/2011/07/tales_from_gdp.html.

On restructuring productivity, see Robert J. Gordon, "Revisiting U.S. Productivity
Growth over the Past Century with a View of the Future," National Bureau for Economic
Research, Working Paper 15834, March 2010. On how firing during the recession relates

to restructuring productivity, see David Berger, "Countercyclical Restructuring and Job-less Recoveries," Yale University, November 16, 2011. On jobless recoveries, see also Nir Jaimovich and Henry E. Siu, "The Trend Is the Cycle: Job Polarization and Jobless Re-coveries," National Bureau of Economic Research, Working Paper, August 2012.

On unfitness to serve in the military, see "A Conversation with Arne Duncan, U.S. Secretary of Education," Council on Foreign Relations, October 19, 2010.

On the cyclical behavior of productivity, see data from the Bureau of Labor Statistics, http://www.bls.gov/news.release/prod2.t01.htm.

On the new class of the long-term unemployed, see Andreas Hornstein, Thomas A. Lubik, and Jessie Romero, "Potential Causes and Implications of the Rise in Long-Term Unemployment," Federal Reserve Bank of Richmond, September 2011, Economic Brief 11-09.

On reservation wages, see Alan B. Krueger and Andreas Mueller, "Job Search and Job Finding in a Period of Mass Unemployment: Evidence from High-Frequency Longitudinal Data," CEPS Working Paper No. 215, January 2011.

On Kaiser, see Sarah Kliff, "Coming to an Insurance Plan Near You: The $32,000 Premium," *The Washington Post*, September 27, 2011. On the median wage for 2010, see the Social Security Administration, http://www.ssa.gov/cgi-bin/netcomp.cgi?year=2010, and Suzy Khimm, "The Median U.S. Wage in 2010 Was Just $26,363," *The Washington Post*, October 20, 2011.

For the story of Rona Economou, see Alex Williams, "Maybe It's Time for Plan C," *The New York Times*, August 12, 2011. For some looks at the recent surge in freelanc-ing, see Sara Horowitz, "The Freelance Surge Is the Industrial Revolution of Our Time," *The Atlantic*, September 1, 2011; Suzy Khimm, "Has the Recession Created a Freelance Utopia or a Freelance Underclass?", *The Washington Post*, September 3, 2011; Ylan Q. Mui, "A Permanent Workforce Shift?: A Surge in Temporary Jobs Reflects a Fundamental Change in American Employment," *The Washington Post*, February 18, 2012; Emily Glazer, "Serfing the Web: Sites Let People Farm Out Their Chores: Workers Choose Jobs, Negotiate Wages; Mr. Kutcher, Anonymously, Asks for Coffee," *The Wall Street Journal*, November 28, 2011; and Jennifer 8. Lee, "Gen-eration Limbo: Waiting It Out," *The New York Times*, August 31, 2011.

For the quotation about Berlin and ambition, see "Berlin's Elections, The Cost of Cool," *The Economist*, September 17, 2011.

Chapter 4: New Work, Old Game

For Gobet on the Herbert Simon quotation, see "Herbert Simon," Chess Program-ming Wiki CPW, http://chessprogramming.wikispaces.com/Herbert+Simon.

Chapter 5: Our Freestyle Future

For the Kasparov quotation, see "Dark horse ZackS wins Freestyle Chess Tournament," *ChessBase News*, June 19, 2005, http://chessbase.com/newsdetail.asp ?new sid=2461, which is also the source for the information on the 2005 Freestyle tournaments.

For some information on Anson Williams, see Daaim Shabazz, "Anson Williams . . . King of Freestyle Chess," http://www.thechessdrum.net/blog/2007/12/21/ anson-williams-king-of-freestyle-chess/, in addition to my interview with him. The Nelson Hernandez quotation comes from the same source. By the way, in Freestyle chess, Anson Williams and Nelson Hernandez have been part of a team for years, but they have never met, instead using the internet and Skype.

For estimates on the strength of some Freestyle teams over the machines, see Vasik Rajlich, "Interviews with Freestylers," http://www.rybkachess.com/docs/free stylers_version_2.htm. The Arno Nickel quotation is from that same source.

The Nakamura quotation is from Arno Nickel, "Freestyle Chess," http://www.free webs.com/freestyle-chess/gmarnonickel.htm.

For a discussion of how opening books work, see this useful piece by Dagh Nielsen, untitled, at http://www.spaghettichess.com/Dagh%20Nielsen_tips.txt.

See http://youtu.be/JSOwlYk_RQU for an Accenture talk by Vishy Anand on finding something new in chess and the importance of memory.

In addition to Freestyle chess there is Correspondence chess. In the old days, pre-computer, chess players frequently played by mail, with lags of two to three days between moves. It was understood that each player would consult books and study the position in depth by moving around the pieces, although it was forbidden to consult other chess players. These days, Correspondence chess is all about the computers. Whereas Freestyle gives you an hour or two for the game, Correspondence chess gives you and the computer a day or more to calculate the correct move. It might seem like Correspondence games should be a lot stronger than Freestyle games, but they're not. The decision tree for chess branches very rapidly (looking fifteen moves ahead involves many more possibilities than searching all possibilities for two moves ahead), so giving the programs a lot more time doesn't make their decisions so much better. In fact Anson Williams believes that the Correspondence form gives the human member of the Freestyle team more data to worry about and may increase some kinds of errors. Vasik Rajlich worries that the humans in Correspondence teams are more likely to think they can tamper with the judgments of the computer, and computer chess expert Ken Regan is actually more impressed by the performances of the Freestyle teams. Even if the extra time with the computer pays off, a Correspondence match may simply boil down to which player doesn't have to hold down a job. It's another form of man–machine collaboration, but for competitive reasons Freestyle chess so far has been more interesting.

For brief overviews of medical diagnoses and AI, see Christopher de la Torre, "The AI Doctor is Ready to See You," *Singularity Hub*, May 10, 2010, http://singularity hub.com/2010/05/10/the-ai-doctor-is-ready-to-see-you/, and Katie Hafner, "For Second Opinion, Consult a Computer?" *The New York Times*, December 3, 2012. One classic treatment is Igor Kononenko, "Machine Learning for Medical Diagnosis: History, State of the Art and Perspective," *Artificial Intelligence in Medicine*, 2001, 23(1): 89–109. On the doubling of the medical literature, see http://jama evidence.com/resource/foreword/520.

For a look at Google for medical diagnostics, see H. Tang and J.H. Ng, "Googling for a Diagnosis—Use of Google as a Diagnostic Aid: Internet Based Study," *BMJ*, December 2, 2006, 333(7579): 1143–45, http://www.ncbi.nlm.nih.gov/pubmed /17098763.

Chapter 6: Why Intuition Isn't Helping You Get a Job

One very good and serious paper on online dating is Eli J. Finkel, Paul W. Eastwick, Benjamin R. Karney, Harry T. Reis, and Susan Sprecher, "Online Dating: A Critical Analysis from the Perspective of Psychological Science," *Psychological Science in the Public Interest*, January 2012, 13(1): 3–66.

For the tale of Cambry, see David Gelles, "Inside Match.com," *Financial Times*, July 29, 2011; this source also has the information on conservatives and liberals and the New Jersey anecdote.

For cognitive biases, see http://en.wikipedia.org/wiki/List_of_cognitive_biases.

For the pointer about experimental economics I am indebted to Amihai Glazer.

In addition to Ken Regan, for another look at using computers to measure the quality of human play, see Matej Guid, "Search and Knowledge for Human and Machine Problem Solving," doctoral dissertation, University of Ljubljana, 2010, http:// eprints.fri.uni-lj.si/1113/1/Matej__Guid.disertacija.pdf. And for a summary of related work, see Matej Guid and Ivan Bratko, "Using Chess Engines to Estimate Human Skill," *Chessbase News*, November 11, 2011, http://www.chessbase.com /newsdetail.asp?newsid=7621. See also the short article "Chess Players Whose Moves Most Matched Computers," *The New York Times*, March 19, 2012.

For a general look at the cognitive abilities of chess players, see Fernand Gobet and Neil Charness, "Expertise in Chess," in *The Cambridge Handbook of Expertise and Expert Performance*, edited by K. Anders Ericsson, Neil Charness, Paul J. Feltovich, and Robert R. Hoffman (New York: Cambridge University Press, 2006): 523–38.

See also Fernand Gobet, Alex de Voogt, and Jean Retschitzki, *Moves in Mind: The Psychology of Board Games* (New York: Psychology Press, 2004).

On males vs. females in chess, see Patrik Gränsmark, *Essays on Economic Behavior, Gender and Strategic Learning* (Stockholm: The Swedish Institute for Social Research, Stockholm University, 2010) for a compilation of these essays. Essay by essay, you can find them separately online. The information is Patrik Gränsmark and Christer Gerdes, "Strategic Behavior Across Gender: A Comparison of Female and Male Expert Chess Players," *Labour Economics*, 2010, 17(5): 766–75; Patrik Gränsmark, "A Rib Less Makes You Consistent but Impatient: A Gender Comparison of Expert Chess Players," Working Paper Series, Swedish Institute for Social Research, May 25, 2010; and Anna Dreber, Christer Gerdes, and Patrik Gränsmark, "Beauty Queens and Battling Knights: Risk Taking and Attractiveness in Chess," Discussion Paper No. 5314, The Institute for the Study of Labor, Bonn, November 2010.

On the strength of previous chess players, see Kenneth W. Regan and Guy McC. Haworth, "Intrinsic Chess Ratings," May, 18, 2011, http://www.cse.buffalo.edu/~regan/papers/pdf/ReHa11c.pdf.

On chess and cognitive improvement and for a survey of the literature, see Robert W. Howard, "Searching the Real World for Signs of Rising Population Intelligence," *Personality and Individual Differences*, April 2001, 30(6): 1039–1058. In addition to Ken Regan's work, one look at the playing strength of previous world champions is Matej Guid, Aritz Pérez, and Ivan Bratko, "How Trustworthy Is CRAFTY's Analysis of World Chess Champions?" *ICGA Journal*, September 2008, 31(3): 131–44.

On women doing better in chess, see NotoriousLTP, "Participation Explains Gender Differences in the Proportion of Chess Grandmasters," *ScienceBlogs*, January 30, 2007, http://scienceblogs.com/purepedantry/2007/01/30/participation-explains-differe/, and C.F. Chabris and M.E. Glickman, "Sex Differences in Intellectual Performance: Analysis of a Large Cohort of Competitive Chess Players," *Psychological Science*, December 2006, 17(12): 1040–46.

For the quotations on looking at all chess through the eyes of the computer, see D. T. Max, "The Prince's Gambit: A Chess Star Emerges for the Post-Computer Age," *The New Yorker*, March 21, 2011.

Chapter 7: The New Office: Regular, Stupid, and Frustrating

For various reports on the failures of GPS, see Tom Vanderbilt, "It Wasn't Me, Officer! It Was My GPS: What Happens When We Blame Our Navigation Devices for Our Car Crashes," *Slate*, June 9, 2010. Ari N. Schulman considers some relevant issues in his "GPS and the End of the Road," *The New Atlantis*, Spring 2011. For a more formal look at some of the problems drivers have with GPS, see Barry Brown and Eric Laurier, "The Normal, Natural Troubles of Driving with GPS,"

CHI 2012, Proceedings of the SIGCHI Conference on Human Factors in Computing Systems, Austin, Texas.

On the difficulties with chess robots, see Duncan Graham-Rowe, "Chess Robots Have Trouble Grasping the Game," *New Scientist*, December 20, 2011, http://www.newscientist.com/blogs/onepercent/2011/12/chess-robots-have-trouble-gras.html.

For a speculative look at other means of applying chess ratings to real-world performance, see Garth Zietsman, "Chess, Intelligence and Winning Arguments," *FreakoStats*, March 16, 2012, http://garthzietsman.blogspot.com/2012/03/chess-intelligence-and-winning.html.

On the Medication Adherence Score, see Scott Thurm, "Next Frontier in Credit Scores: Predicting Personal Behavior," *The Wall Street Journal*, October 27, 2011, which is also a source for the data on credit checks.

The Vonnegut quotation is from Kurt Vonnegut, *Player Piano* (New York: Dial Press, 2006), p. 93, originally published 1952. See also p. 161 for a discussion of how the machines can accurately evaluate human talent.

Chapter 8: Why the Turing Game Doesn't Matter

For one look at the views of Eliezer Yudkowsky, see Robin Hanson, "Debating Yudkowsky," *Overcoming Bias*, July 3, 2011, http://www.overcomingbias.com/2011/07/debating-yudkowsky.html.

On the difficulties of building full AI without accompanying bodies, see Virginia Hughes, "Body Conscious: When It Comes to Artificial Intelligence, the Brain Isn't Everything," *New Scientist*, August 20, 2011. See also David J. Linden, "The Singularity Is Far: A Neuroscientist's View," *BoingBoing*, July 14, 2011, http://boingboing.net/2011/07/14/far.html. Another useful essay is David Robson, "Your Clever Body: Thinking from Head to Toe," *New Scientist*, October 15, 2011.

For the dialogue with Rosette, see http://orgtheory.wordpress.com/2011/10/22/2011-loebner-prize-artificial-intelligence-still-has-a-long-way-to-go/.

For a Tyler Cowen and Michelle Dawson paper on Turing, see "What Does the Turing Test Really Mean? And How Many Human Beings (Including Turing) Could Pass?", June 3, 2009, http://www.gmu.edu/centers/publicchoice/faculty%20pages/Tyler/turingfinal.pdf. The characterization of Turing in particular, and his views in his famous essay, is taken from the article, although Dawson is not liable for any mistakes or shortcomings in this particular use of the material.

You can find the dialogue with Cleverbot here: http://www.geekologie.com/2009/02/cleverbot-arguably-clever-want.php.

For the studies of which online approaches work, see Valentin Schöndienst and Linh Dang-Xuan, "The Role of Linguistic Properties in Online Dating Communication—A Large-Scale Study of Contact Initiation Messages," http://www.pacis-net.org/file/2011/PACIS2011-166.pdf.

The two sketches are taken from Torie Bosch, "How Robots Saved an Artist's Sanity: The Greatest Artist of His Generation Is Named Paul." *Slate*, November 15, 2012. On the aesthetics of chess, see Azlan Iqbal (with Harold van der Heijden, Matej Guid and Ali Makhmali), "A Computer Program to Identify Beauty in Problems and Studies," *Chessbase News*, December 15, 2012, http://www.chessbase.com/news detail.asp?newsid=8602.

On the chess cheating scandal, see for instance Henry Samuel, "Chess World Rocked by French Cheating Scandal," *The Telegraph*, March 25, 2011, and also "French Chess Federation Suspends Players Accused of Cheating," *ChessBase News*, March 21, 2011, http://www.chessbase.com/newsdetail.asp?newsid=7094. For a more general look at Ken Regan's study of computer cheating in chess, see Dylan Loeb McClain, "To Detect Cheating in Chess, a Professor Builds a Better Program," *The New York Times*, March 19, 2012. Cheating doesn't have to involve computers, machines, and phony compartments with humans inside. Charles Moul and John Nye, both economists, did a study of Soviet chess players from 1940 to 1964. American grandmaster Bobby Fischer notoriously had charged that the Soviets cheated during those years. He believed that they would throw games to favored players, agree to easy draws to conserve strength when needed in long tournaments, and in general manipulate tournament results to stop any non-Soviet player from taking first place. Moul and Nye went back to the games and found that the charge of cheating was true, in particular with regard to collusion through draws. Fischer had been right and the Soviets had been cheating all along. Charles C. Moul and John V. Nye, "Did the Soviets Collude? A Statistical Analysis of Championship Chess 1940–64," *Journal of Economic Behavior and Organization*, 2009, 70(1-2): 10–21.

On using the Turk to feign a Turing test, see Tom Standage, *The Mechanical Turk: The True Story of the Chess-Playing Machine That Fooled the World* (London: The Penguin Press, 2002), p. 215. The book is also a good introduction to the history of the Turk. On the talking machine, see Bradley Ewart, *Chess: Man vs Machine* (San Diego: A.S. Barnes & Company, 1980), pp. 26–28. The book in general has a great deal of information about the Mechanical Turk and its history and operation.

On Google and memory, see Betsy Sparrow, Jenny Liu, and Daniel M. Wegner, "Google Effects on Memory: Cognitive Consequences of Having Information at Our Fingertips," *Science*, August 5, 2011.

On memory and innovation, see for instance "General Introduction," by Mary Carruthers and Jan M. Ziolkowski, in *The Medieval Craft of Memory: An Anthology of Texts and Pictures*, edited by Mary Carruthers and Jan M. Ziolkowski (Philadelphia: University of Pennsylvania Press, 2002), in particular pp. 3–4.

For a brief report on Martin Thoresen and his tournaments, see Dylan Loeb McClain, "The New Computer Chess Bully on the Screen," *The New York Times*, Gambit, the Chess Blog, January 11, 2011, http://gambit.blogs.nytimes.com/tag/martin-thoresen/.

On overconfidence against the Mechanical Turk, see Bradley Ewart, *Chess: Man vs Machine* (San Diego: A.S. Barnes & Company, 1980), p. 76. In the field of neuroeconomics, Nobel laureate Vernon Smith and his coauthors did a famous study of how cooperators respond differently when they know they are facing humans rather than computers. See Kevin McCabe, Daniel Houser, Lee Ryan, Vernon Smith, and Theodore Trouard, "A Functional Imaging Study of Cooperation in Two-Person Reciprocal Exchange," *PNAS*, September 25, 2001, 98(20): 11,832–35, doi: 10.1073/pnas.211415698.

Chapter 9: The New Geography

On labor's falling share in output, see Florence Jaumotte and Irina Tytell, "How Has the Globalization of Labor Affected the Labor Income Share in Advanced Countries?", IMF Working Paper, 2007.

For a look at what is in the textbook of Borjas, on immigration, see Bryan Caplan, "Borjas, Wages, and Immigration: The Complete Story," *EconLog*, March 16, 2007, http://econlog.econlib.org/archives/2007/03/borjas_wages_an.html.

For the most comprehensive theoretical look at offshoring issues, see Daron Acemoglu, Gino Gancia, and Fabrizio Zilibotti, "Offshoring and Directed Technical Change," National Bureau of Economic Research, Working Paper 18595, December 2012.

On multinational job creation and destruction, see Jia Lynn Yang, "Corporations Pushing for Job-Creation Tax Breaks Shield U.S.-vs.-Abroad Hiring Data," *The Washington Post*, August 21, 2011.

For Autor's work on offshoring, see David H. Autor, David Dorn, and Gordon H. Hanson, "The China Syndrome: Local Labor Market Effects of Import Competition in the United States," National Bureau of Economic Research, Working Paper 18054, May 2012. Also relevant is Runjuan Liu and Daniel Trefler, "A Sorted Tale of Globalization: White Collar Jobs and the Rise of Service Offshoring," National Bureau of Economic Research, Working Paper 17559, November 2011, and that is the source for information on downward occupational switching.

On how much of the US economy comes from China, see Galina Hale and Bart Hobijn, "The U.S. Content of 'Made in China'," Federal Reserve Board of San Francisco Economic Letter, August 8, 2011.

For an overview of the research on the connection between immigration and offshoring, see Tyler Cowen, "How Immigrants Create More Jobs," *The New York Times*, October 30, 2010.

On the importance of economic clustering, see the blog post by Noah Smith, "Great Stagnation . . . or Great Relocation?", http://noahpinionblog.blogspot.com/2011/09 /great-stagnationor-great-relocation.html.

On how cities are diverging, see Sabrina Tavernise, "A Gap in College Graduates Leaves Some Cities Behind," *The New York Times*, May 30, 2012. Especially on convergence, see also Peter Ganong and Daniel Shoag, "Why Has Regional Convergence in the U.S. Stopped?" SSRN working paper, March 28, 2013, and also Enrico Moretti, *The New Geography of Jobs* (Boston: Houghton Mifflin Harcourt, 2012).

On the geographic concentration of wage benefits from the internet, see Chris Forman, Avi Goldfarb, and Shane Greenstein, "The Internet and Local Wages: A Puzzle," *American Economic Review*, Febuary 2012, 102(1): 556–75.

On the German population growing again, see Suzanne Daley and Nicholas Kulish, "Brain Drain Feared as German Jobs Lure Southern Europeans," *The New York Times*, April 28, 2012.

On job growth in services, see A. Michael Spence and Sandile Hlatshwayo, "The Evolving Structure of the American Economy and the Employment Challenge," Council on Foreign Relations, March 2011.

For one recent look at reshoring, see John Markoff, "Skilled Work, Without the Worker," *The New York Times*, August 18, 2012.

Chapter 10: Relearning Education

For figures on K–12, see Stephanie Banchero and Stephanie Simon, "My Teacher is an App," *The Wall Street Journal*, November 12, 2011.

The point about stronger incentives for innovation I owe to Alex Tabarrok.

On the Emporium model, see Daniel de Vise, "At Virginia Tech, computers help solve a math class problem," *The Washington Post*, April 22, 2012.

On spelling bees, see Angela Lee Duckworth, Teri A. Kirby, Eli Tsukayama, Heather Berstein, and K. Anders Ericsson, "Deliberate Practice Spells Success: Why Grittier Competitors Triumph at the National Spelling Bee," *Social Psychological and Personality Science*, published online October 4, 2010, doi: 10.1177/1948550610385872.

On Jesse Kraai, see Scott Kraft, "Chess Players Making Right Moves at Younger Ages," the *Los Angeles Times*, May 10, 2011.

On more general progress with prodigies, see Jonathan Wai, Martha Putallaz, and Matthew C. Makel, "Studying Intellectual Outliers: Are There Sex Differences, and

Are the Smart Getting Smarter?" *Current Directions in Psychological Science*, December 2012, 21(6): 382–90, doi:10.1177/0963721412455052.

For two studies of KIPP, see Joshua D. Angrist, Susan M. Dynarski, Thomas J. Kane, Parag A. Pathak, and Christopher R. Walters, "Who Benefits From KIPP?" National Bureau of Economic Research, Working Paper 15740, February 2010; and Christina Clark Tuttle, Bing-ru Teh, Ira Nichols-Barrer, Brian P. Gill, and Philip Gleason, "Student Characteristics and Achievement in 22 KIPP Middle Schools," unpublished manuscript, June 2010.

The Amanda Ripley quotation is from Amanda Ripley, "Boot Camp for Teachers," *The Atlantic*, July/August 2012.

On Hong Kong celebrity tutors, see Hillary Brenhouse, "Meet the Glamorous Celebrity Tutors of Hong Kong," *Slate*, August 29, 2011.

For one look at the value of conscientiousness, see Angela L. Duckworth, David Weir, Eli Tsukayama, and David Kwok, "Who Does Well in Life? Conscientious Adults Excel in both Objective and Subjective Success," *Frontiers in Psychology*, September 2012, 3:356. James J. Heckman considers the relevance of various personality traits in his "Integrating Personality Psychology into Economics," National Bureau of Economic Research, Working Paper, 17378, August 2011.

Chapter 11: The End of Average Science

On the history of proofs and their ambiguities, see for instance Dick Lipton, "Deolalikar's Claim: One Year Later," August 11, 2011, http://rjlipton.wordpress.com/2011/08/11/deolalikars-claim-one-year-later/.

On coauthored pieces in economics, see Daniel S. Hamermesh, "Six Decades of Top Economics Publishing: Who and How?" National Bureau of Economic Research, Working Paper 18635, December 2012. That same piece is also the source on publishing economists growing older and also relying more on originally created data sets.

I accessed the Wikipedia entry on string theory on December 26, 2012. Perhaps it will become clearer!

On the age dynamics for achievement for non-economists, see Benjamin F. Jones and Bruce A. Weinberg, "Age Dynamics in Scientific Creativity," published online before print, *PNAS*, November 7, 2011, doi: 10.1073/pnas.1102895108.

On data crunching pushing out theory, see the famous essay by Leo Breiman, "Statistical Modeling: The Two Cultures," *Statistical Science*, 2001, 16(3): 199–231, including the comments on the piece as well. See also the recent piece by Betsey Stevenson and Justin

Wolfers, "Business is Booming in Empirical Economics," Bloomberg.com, August 6, 2012. And as mentioned earlier, see Daniel S. Hamermesh, "Six Decades of Top Economics Publishing: Who and How?" National Bureau of Economic Research, Working Paper 18635, December 2012. Peter Norvig on the cultures of statistics and statistical learning is another source of relevance; see for instance Peter Norvig, "On Chomsky and the Two Cultures of Statistical Learning," http://norvig.com/chomsky.html.

Chapter 12: A New Social Contract?

The Congressional Budget Office has estimated that just closing our current budget deficit on tax increases alone would require doubling tax rates. See "The Long-Term Effects of Some Alternative Budget Policies," Congressional Budget Office, May 19, 2008, http://www.cbo.gov/ftpdocs/92xx/doc9216/Letter-to-Ryan.1.1.shtml.

On migration, including migration into Texas, see http://www.migration information.org/datahub/state.cfm?ID=TX#table3. For related background, see also Peter Ganong and Daniel Shoag, "Why Has Regional Income Convergence in the U.S. Stopped?", SSRN working paper, March 28, 2013. Other sources on migration and outmigration, include http://www.census.gov/prod/2006pubs/p25-1135.pdf.

On Texas weather, see Ana Campoy, "Heat Scorches Parched Texas," *The Wall Street Journal*, August 6, 2011.

On home prices in Houston and Brooklyn, see Kevin D. Williamson, "Paul Krugman is Still Wrong About Texas," *National Review Online*, August 15, 2011.

For various data on recipients of Medicare, see Kaiser Family Foundation, "Projecting Income and Assets: What Might the Future Hold for the Next Generation of Medicare Beneficiaries?", June 2011, http://www.kff.org/medicare/8172.cfm, and on median net worth see http://www.census.gov/prod/2008pubs/p70-115.pdf, Table 4.

The Knoxville information can be sampled on Apartments.com and Apartment Guide.com.

On El Paso and Juarez, see Andrew Rice, "Life on the Line," *The New York Times Magazine*, July 31, 2011.

On the vulnerability of women, see Gregory Acs, "Downward Mobility from the Middle Class: Waking up from the American Dream," Pew Charitable Trusts, Economic Mobility Project, 2011. That is also the source for the insights about individuals falling out of the middle class.

The Richard Florida quotation is from Richard Florida, "The Conservative States of America," *The Atlantic*, March 29, 2011.

Acknowledgments

For useful discussions and comments I wish to thank Nelson Hernandez, Anson Williams, Kenneth Regan, Jason Fichtner, Erik Brynolfsson, Andrew McGee, Don Peck, Derek Thompson, Michelle Dawson, Peter Snow, Veronique de Rugy, Garett Jones, Robin Hanson, Bryan Caplan, Alex Tabarrok, Natasha Cowen, Garry Kasparov, Vasik Rajlich, Stephen Morrow, David Brooks, Peter Thiel, Michael Mandel, and Larry Kaufman, with apologies to anyone I may have left out.

Index

Also by Tyler Cowen

978-0-452-29884-2

978-0-525-95271-8

978-0-452-28963-5

DISCOVER YOUR
INNER ECONOMIST

Use Incentives to
Fall in Love, Survive Your Next Meeting,
and Motivate Your Dentist

TYLER COWEN

www.marginalrevolution.com

978-0-452-29619-0